Culinary Nutrition

Principles and Applications

STUDY GUIDE

atp
AMERICAN TECHNICAL PUBLISHERS
Orland Park, Illinois 60467-5756

Linda J. Trakselis, MS
Eric M. Stein, MS, RD, CCE

The *Culinary Nutrition Principles and Applications Study Guide* contains procedures commonly practiced in the foodservice industry. Specific procedures vary with each task and must be performed by a qualified person. For maximum safety, always refer to specific manufacturer recommendations, insurance regulations, specific job site procedures, applicable federal, state, and local regulations, and any authority having jurisdiction. The material contained is intended to be an educational resource for the user. American Technical Publishers, Inc. assumes no responsibility or liability in connection with this material or its use by any individual or organization.

1 2 3 4 5 6 7 8 9 – 14 – 9 8 7 6 5 4 3 2 1

Printed in the United States of America

ISBN 978-0-8269-4222-7

This book is printed on recycled paper.

CONTENTS

Shaping Eating Behaviors

Chapter 1

REVIEW

Name ______________________________ **Date** ____________________

True-False

T F **1.** Heat cannot increase the nutrient value of foods.

T F **2.** Layers of flavors are created when different ingredients are added at different times during the preparation and cooking processes.

T F **3.** Processed foods containing high amounts of sugars, fats, and/or sodium are often less expensive than foods containing quality fresh ingredients.

T F **4.** Menu items will only sell on the basis that they are "healthy."

T F **5.** Oil is derived from vegetables, nuts, seeds, and fruits.

T F **6.** Research supports the idea that tastes can be categorized as sweet, salty, sour, bitter, or umami.

T F **7.** Temperature affects the aroma of food.

T F **8.** Bones are never used as a flavoring component in meat, poultry, or fish stocks.

T F **9.** Spices are always added at the end of the cooking process.

T F **10.** Heat typically flows from a cooler material to a warmer material.

T F **11.** A gum is a thickening agent derived from plant flours.

T F **12.** A rub is a blend of ingredients that is pressed onto the surface of uncooked foods to impart flavor.

T F **13.** Sustainable agriculture is a method of food production that seeks to conserve land and water, reduce energy consumption, avoid the use of harmful fertilizers and pesticides, and limit air pollution.

T F **14.** The Maillard reaction occurs when the lipids and sugars in a food are exposed to heat and merge together to form a brown exterior.

T F **15.** Dry-heat cooking is any cooking technique that uses hot air, hot metal, a flame, or hot fat to conduct heat and brown food.

Multiple Choice

__________ **1.** According to the Food and Drug Administration (FDA), milk, eggs, fish, crustacean shellfish, tree nuts, peanuts, wheat, and ___ account for 90% of all food allergies.
- A. soybeans
- B. rice
- C. corn
- D. quinoa

__________ **2.** The Nutrition Facts label provides a standard for comparing foods based on serving size, calories, nutrients, and percent daily value for a ___ calorie diet.
- A. 1000
- B. 1500
- C. 2000
- D. 2500

__________ **3.** A ___ is an individual who primarily consumes plant-based foods and only consumes animal-based foods, such as meat, dairy, and eggs, occasionally or in small amounts.
- A. flexitarian
- B. lacto vegetarian
- C. vegetarian
- D. vegan

__________ **4.** A(n) ___ food is a food that has been altered by processes such as canning, cooking, freezing, dehydration, or milling.
- A. whole
- B. raw
- C. organic
- D. processed

__________ **5.** Standard nutrition labels are required to include ___.
- A. unsaturated fat and "trans" fat
- B. saturated fat and unsaturated fat
- C. total grams of fat, saturated fat, and unsaturated fat
- D. total grams of fat, saturated fat, and "trans" fat.

__________ **6.** Sour tastes come from the ___ in foods.
- A. glutamate
- B. salts
- C. acids
- D. alkalies

__________ **7.** Vinegar is a sour, acidic liquid made from fermented ___.
- A. alcohol
- B. oil
- C. seeds
- D. herbs

______________ **8.** Caramelization is a reaction that occurs when ___ are exposed to high heat and results in browning and a change in flavor.
A. salts
B. sugars
C. lipids
D. proteins

______________ **9.** A percent daily value of 5% or less is considered low, whereas a percent daily value of ___% or more is regarded as high.
A. 10
B. 20
C. 30
D. 40

______________ **10.** An herb is a flavoring derived from the leaves or ___ of a very aromatic plant.
A. berries
B. roots
C. bark
D. stem

______________ **11.** Meat, poultry, or seafood that is labeled "lean" has less than ___ g of saturated fat per 100 g.
A. 4
B. 5
C. 6
D. 7

______________ **12.** The percentages of vitamin A, vitamin C, ___, and ___ are required on standard Nutrition Facts labels.
A. vitamin D; vitamin K
B. potassium; sodium
C. calcium; iron
D. chloride; zinc

______________ **13.** A ___ is a thick liquid with a flavor that has been concentrated as a result of being gently simmered until reduced in volume.
A. condiment
B. marinade
C. stock
D. reduction

______________ **14.** ___ have an umami taste.
A. Mushrooms, tomatoes, and cheeses
B. Spinach, chocolate, and coffee
C. Sugar and milk
D. Citrus fruits and vinegar

____________________ **15.** A(n) ___ is a flavoring derived from the bark, seeds, roots, flowers, berries, or beans of an aromatic plant.

A. herb
B. spice
C. extract
D. alcohol

____________________ **16.** A ___ is a flavorful liquid used to soak uncooked foods to impart flavor and sometimes tenderize.

A. chutney
B. mirepoix
C. marinade
D. dry rub

____________________ **17.** Cheeses, fruits, and wines generally have more developed flavors at ___ temperature.

A. a cold
B. a hot
C. a freezing
D. room

____________________ **18.** A ___ is a sauce typically made from either raw or cooked puréed fruits or vegetables.

A. coulis
B. chutney
C. béchamel
D. hollandaise

____________________ **19.** ___ is a heat transfer method in which heat passes from one object to another through direct contact.

A. Conduction
B. Convection
C. Infrared radiation
D. Induction radiation

____________________ **20.** The cookware used on a(n) ___ cooktop must be flat-bottomed and made of a material that will conduct electricity, such as cast iron or magnetic stainless steel.

A. convective
B. microwave
C. infrared
D. induction

____________________ **21.** Moist-heat cooking techniques include simmering, boiling, blanching, steaming, and ___.

A. poaching
B. smoking
C. stir-frying
D. stewing

_______________ **22.** ___ are the most common seasonings.
A. Salts and peppercorns
B. Herbs and spices
C. Alcohols and extracts
D. Oils and juices

_______________ **23.** A liaison is a thickening agent that is a mixture of ___ used to thicken sauces.
A. flour and fat
B. cornstarch and water
C. egg yolks and heavy cream
D. egg whites and milk

_______________ **24.** If stored properly, fresh herbs will typically remain fresh up to ___ days.
A. five
B. seven
C. nine
D. ten

_______________ **25.** ___ can make sweet foods taste sweeter, make sour foods taste more sour, and intensify the flavor of mild foods.
A. Vinegar
B. Oil
C. Pepper
D. Salt

Completion

_______________ **1.** A(n) ___ is an adverse reaction to a specific food that does not involve the immune system.

_______________ **2.** The senses of taste and ___ are particularly significant because they are the dominant senses used by the brain to detect flavors.

_______________ **3.** ___ is the process of arranging food in an appealing manner for presentation.

_______________ **4.** ___ is the way a food or beverage feels in the mouth.

_______________ **5.** A(n) ___ is a the combination of ingredients, such as herbs, spices, and seasonings, associated with a specific cuisine.

_______________ **6.** ___ is a heat transfer method that occurs due to the circular movement of a fluid or gas.

_______________ **7.** ___ is the Japanese word for "deliciousness" and is a savory taste.

_______________ **8.** ___ is a sense that is activated by the receptor cells that make up the taste buds.

_______________ **9.** The ___ is the temperature at which oil begins to smoke and emit an odor.

_______________ **10.** ___ is the colored, outermost layer of a citrus fruit peel and contains a high concentration of oil.

Critical Thinking

Use complete sentences to respond to the following.

1. Summarize the four major influences on food choices.

2. Describe how flavor is influenced by flavor perceptions, presentation, textures, tastes, aromas, and temperatures.

3. Compare the dry-heat, moist-heat, and combination cooking techniques that are considered healthy.

Activity 5-5: Mouthfeel of Fats and Oils Lab

Fats and oils help disperse flavors throughout the palate by coating the taste buds. Both the amount and type of fat in foods produce different mouthfeels that can be described as light, medium, or heavy.

Objective: To identify the mouthfeel of foods that are key sources of fats and oils.

Materials and Equipment

- Tasting portions of the following foods:
  - avocado
  - canola oil
  - commercially prepared cake frosting
  - olive oil
  - plain yogurt
  - real mayonnaise
  - room temperature butter
  - room temperature coconut oil
  - sour cream
- 9 tasting cups
- 9 tasting spoons

Allergen Alert

The materials used in this activity may contain one or more food allergens.

Procedure

1. Place a tasting portion of each of the nine identified foods in a tasting cup.
2. Taste each of the nine identified foods.
3. After each food is tasted, place an "X" in the chart to identify the mouthfeel of each food.

	Food	Light Coating	Medium Coating	Heavy Coating
1.	Avocado			
2.	Canola oil			
3.	Commercially prepared cake frosting			
4.	Olive oil			
5.	Plain yogurt			
6.	Real mayonnaise			
7.	Room temperature butter			
8.	Room temperature coconut oil			
9.	Sour cream			

Activity 5-4: Lipid Dietary Analysis

If more calories are consumed than are burned through physical activity, body fat will accumulate. Excess levels of body fat put an individual at risk for developing high blood pressure, type 2 diabetes, and heart disease. In contrast, a lipid deficiency interferes with the absorption of the fat-soluble vitamins. Symptoms of a lipid deficiency include the inability to regulate body temperature, muscle weakness, and dry skin, hair, and lips.

Access your online dietary analysis profile that was created in Activity 2-7 from one of the following websites:

FatSecret™: www.fatsecret.com
SelfNutritionData: http://nutritiondata.self.com/mynd/mytracking
SparkPeople®: www.sparkpeople.com
SuperTracker: www.supertracker.usda.gov

1. Submit a copy of your seven-day dietary analysis.
2. Submit a written report that includes the following criteria:
   a. Describe how the daily fat intake compares with USDA dietary recommendations for your age, gender, and physical activity level.
   b. Compare the amount of unsaturated fats consumed with the amount of saturated and trans fats consumed.
   c. Identify three changes that could be made to improve fat consumption.

Dietary Analysis Rubric

Content	
Copy of dietary analysis	10
Written report	5
Quality content	10
Summary	5
Format	
Typed, double spaced	5
Correct spelling	5
Correct grammar	5
Correct punctuation	5
Total	50

Shaping Eating Behaviors

Chapter 1

ACTIVITIES

Name ______________________________ **Date** ______________

Activity 1-1: Food Labels

Food labels contain an ingredient list and Nutrition Facts that can help individuals identify nutrient-dense foods. These labels are also helpful when comparing similar products to determine which product best meets the nutritional criteria of the user.

Choose three products from one of the following categories: cereals, soups, snack foods, or yogurts. Use the food label on each product to complete the following.

1.

Food Label	Name of Product 1	Name of Product 2	Name of Product 3
First three ingredients on ingredient list			
Serving size			
Calories			
Calories from fat			
Total fat (g)			
Saturated fat (g)			
Trans fat (g)			
Cholesterol (mg)			
Sodium (mg)			
Total carbohydrates (g)			
Dietary fiber (g)			
Sugars (g)			
Protein (g)			

Based on one serving size, list the product (1, 2, or 3) that has the lowest amount in each category.

_______________ **2.** Calories

_______________ **3.** Calories from fat

_______________ **4.** Total fat

_______________ **5.** Saturated fat

_______________ **6.** Trans fat

_______________ **7.** Cholesterol

_______________ **8.** Sodium

_______________ **9.** Total carbohydrates

_______________ **10.** Dietary fiber

_______________ **11.** Sugars

_______________ **12.** Protein

13. Which food group do the products listed represent? What is the recommended daily serving size for this food group based on a 2000 calorie diet?

Activity 1-2: Cooking Techniques

Foods are cooked using dry-heat, moist-heat, and combination cooking techniques in order to enhance their flavors, make them easier to digest, and kill any harmful microorganisms that may be present.

Match the cooking technique to its description.

______________ **1.** Food is cooked using direct heat from above.

______________ **2.** Cooking technique commonly used with large cuts of meat or whole poultry where the item is cooked uncovered at a high temperature in an oven.

______________ **3.** Food is cooked in liquid that is between 185°F and 205°F.

______________ **4.** Bite-sized pieces of food are covered with liquid and simmered for a long period of time in a covered pot.

______________ **5.** Food is cooked by completely submerging it in hot oil.

______________ **6.** Food is cooked using direct heat from below to yield a smoky flavor and char marks.

______________ **7.** Food is cooked in liquid with rapidly rising bubbles that break at the surface.

______________ **8.** Food is cooked in a small amount of fat in a wok.

______________ **9.** Food is cooked quickly in a hot pan that is set over direct heat and contains a small amount of oil.

______________ **10.** Cooking technique commonly used with pancakes, eggs, and hot sandwiches where the food is cooked on a solid, flat metal surface.

______________ **11.** Cooking technique commonly used with fabricated poultry, breads, and cakes where the item is cooked with dry hot air in an enclosed environment.

______________ **12.** Food is seared and then cooked, tightly covered, in a small amount of liquid for a long period of time.

______________ **13.** Food is cooked in liquid that is between 160°F and 180°F.

______________ **14.** Food is cooked in a pan containing a medium amount of hot fat.

______________ **15.** Food is cooked in a container that is placed above hot water to prevent steam from escaping.

Dry Heat

A. Broiling
B. Grilling
C. Roasting
D. Baking
E. Griddling
F. Sautéing
G. Stir-frying
H. Pan-frying
I. Deep-frying

Moist-Heat

J. Poaching
K. Simmering
L. Boiling
M. Steaming

Combination

N. Braising
O. Stewing

Activity 1-3: Fresh Herbs and Dried Spices Lab

Herbs and spices have unique appearances, aromas, and flavors that help develop flavor in both savory and sweet recipes. Herbs and spices are virtually calorie-free and their use does not add sugar, fat, or sodium to menu items.

Objective: To identify herbs and spices based on appearance, aroma, and flavor.

Materials and Equipment

- Marker
- 10 tasting cups
- Fresh herbs
  - basil
  - chives
  - cilantro
  - rosemary
  - thyme
- Blindfold
- Teaspoon
- Dried ground spices
  - cardamom
  - cumin
  - dry mustard
  - nutmeg
  - turmeric
- 10 tasting spoons

Procedure

Fresh herbs

1. Label the bottoms of five of the tasting cups, one herb per label.
2. Place a sample of each herb into the correctly labeled cup.
3. Working in pairs, take turns placing the cups in random order and identifying each herb based on its appearance.
4. Place a blindfold on one partner and have that partner smell and identify each herb based on its aroma.
5. With the blindfold still in place, have the partner taste and identify each herb based on its flavor.
6. Repeat steps 4–5 with the other lab partner wearing the blindfold.

Dried spices

7. Label the bottoms of the five remaining tasting cups, one spice per label.
8. Place 1 tsp of each spice into the correctly labeled cup.
9. Working in pairs, take turns placing the cups of spices in random order and identifying each spice based on its appearance.
10. Place a blindfold on one partner and have that partner smell and identify each spice based on its aroma.
11. With the blindfold still in place, have the partner taste and identify each spice based on its flavor.
12. Repeat steps 10–11 with the other lab partner wearing the blindfold.

Scoring: A point is earned for each herb and spice correctly identified based on its appearance, aroma, and flavor for a possible total score of 30 points.

Activity 1-4: Oils and Vinegars Lab

Oils and vinegars are staples in the professional kitchen and provide a vast range of flavors. Oils supply heart-healthy fats, while vinegars are fat-free, low in sodium and calories, and mostly low in sugar.

Objective: To identify oils based on aroma and flavor and to identify vinegars based on appearance, aroma, and flavor.

Materials and Equipment

- Marker
- 10 tasting cups
- Oils
  - canola oil
  - corn oil
  - extra-virgin olive oil
  - peanut oil
  - sesame oil
- Blindfold
- Tablespoon
- 20 tasting spoons
- Vinegars
  - balsamic vinegar
  - cider vinegar
  - red wine vinegar
  - sherry vinegar
  - rice wine vinegar

Allergen Alert

The materials used in this activity may contain one or more food allergens.

Procedure

Oils

1. Label the bottoms of five of the tasting cups, one type of oil per label.
2. Place 1 tbsp of each oil into the correctly labeled cup.
3. Place a blindfold on one partner and have that partner smell and identify each type of oil based on its aroma.
4. With the blindfold still in place, have the partner taste and identify each type of oil based on its flavor.
5. Repeat steps 3–4 with the other lab partner wearing the blindfold.

Vinegars

6. Label the bottoms of the five remaining tasting cups, one type of vinegar per label.
7. Place 1 tbsp of each vinegar into the correctly labeled cup.
8. Working in pairs, take turns placing the cups of vinegar in random order and identifying each type of vinegar based on its appearance.
9. Place a blindfold on one partner and have that partner smell and identify each type of vinegar based on its aroma.
10. With the blindfold still in place, have the partner taste and identify each type of vinegar based on its flavor.
11. Repeat steps 9–10 with the other lab partner wearing the blindfold.

Scoring: A point is earned for each oil correctly identified based on its aroma and flavor and a point for each vinegar correctly identified based on its appearance, aroma, and flavor for a possible total score of 25 points.

Activity 1-5: Sustainable Farms Research

Sustainable agriculture is a method of food production that seeks to conserve land and water, reduce energy consumption, avoid the use of harmful fertilizers and pesticides, and limit air pollution.

Access the Eat Well Guide® at www.eatwellguide.org or a website of choice and choose a farm that practices sustainable agriculture. Prepare a three- to four-page report and an oral presentation that addresses the following criteria.

1. Describe three sustainable practices used on the farm.
2. Explain why each of these practices are used.
3. Identify two pros and two cons of using sustainable practices.
4. Discuss the impact of sustainable agriculture on future farming in terms of practicality, expense, time, production, and yield.
5. Discuss at least three nutritional implications for the products produced by the farm.

Nutrition Project Rubric

Content	
Written report	5
Cover page	5
Introduction	5
Quality content	20
Summary	5
Format	
Typed, double spaced	5
Correct spelling	5
Correct grammar	5
Correct punctuation	5
Bibliography	5
Oral Presentation	
Introduction	5
Quality content	10
Summary	5
Delivery	5
Visual aid(s)	10
Total	100

Nutrition Fundamentals Chapter 2 REVIEW

Name ______________________________ Date ____________________

True-False

T F **1.** Cardiovascular disease is the primary cause of death in the United States.

T F **2.** In order to maintain weight, the calories consumed and calories used by the body need to be equal.

T F **3.** According to the CDC, individuals eat up to 5% more food when presented with larger portions.

T F **4.** Water, vitamins, and minerals supply energy but do not have calories.

T F **5.** The USDA recommends that individuals keep fat consumption in a range of 10–14 tsp daily.

T F **6.** The body burns calories through movement and physical activity.

T F **7.** The purpose of the culinary nutrition recipe-modification process is to change an existing recipe into a healthier one.

T F **8.** According to the *2010 Dietary Guidelines for Americans,* solid fats should replace oils when possible.

T F **9.** After identifying the ingredients to be modified, the next step in the culinary nutrition recipe-modification process is to evaluate the original recipe.

T F **10.** A nutrient-dense food is a food that is low in nutrients and high in calories.

T F **11.** The body mass index does not account for muscle mass, which weighs more than fat.

T F **12.** Serving sizes differ for each food group because each group provides different amounts of calories and nutrients.

T F **13.** Most animal-based proteins provide a significant amount of dietary fiber.

T F **14.** The American Heart Association recommends limiting daily sodium intake to less than 1500 mg per day.

T F **15.** When modifying a recipe, an appealing final product can often be produced simply by reducing or eliminating the identified ingredient.

Multiple Choice

______________ **1.** Proteins, carbohydrates, lipids, and vitamins are ___ nutrients.
A. macro-
B. micro-
C. organic
D. inorganic

______________ **2.** Gastric juices begin the process of digestion ___ by separating them into their most basic forms.
A. carbohydrates and proteins
B. carbohydrates and lipids
C. proteins and lipids
D. carbohydrates, proteins, and lipids

______________ **3.** Proteins and carbohydrates both provide ___ calories per gram.
A. 2
B. 4
C. 6
D. 8

______________ **4.** A(n) ___ food is a food that is high in calories and low in nutrients.
A. nutrient-dense
B. functional
C. fortified
D. empty-calorie

______________ **5.** A ___ is an energy-providing nutrient made of carbon, hydrogen, oxygen, and nitrogen assembled in chains of amino acids.
A. protein
B. carbohydrate
C. mineral
D. vitamin

______________ **6.** The portal system is a system for transporting water-soluble nutrients from the small intestine to the ___ and then into the bloodstream.
A. pancreas
B. gallbladder
C. large intestine
D. liver

______________ **7.** In the Harvard Healthy Eating Plate, fish, poultry, beans, and ___ are stressed as protein sources.
A. nuts
B. red meats
C. processed meats
D. refined grains

_______________ **8.** The ___ serves as the primary site for nutrient absorption.
A. mouth
B. stomach
C. small intestine
D. large intestine

_______________ **9.** A(n) ___ is the estimated daily intake level for a nutrient for which no RDA has been set but nutrient intake appears to be adequate to meet the needs of a healthy population.
A. estimated average requirement (EAR)
B. recommended dietary allowance (RDA)
C. tolerable upper intake level (UL)
D. adequate intake (AI)

_______________ **10.** The daily recommended serving size for vegetables ranges from ___ cups depending on age, gender, and level of physical activity.
A. 1–2
B. 1–3
C. 3–4
D. 3–5

_______________ **11.** Hypertension is a condition characterized by ___.
A. elevated blood sugar levels
B. high blood pressure
C. abnormal cell growth
D. rapid heartbeat

_______________ **12.** Celiac disease is a condition in which ___ damages the ability of the small intestine to absorb nutrients.
A. gluten
B. lactose
C. soy
D. cheese

_______________ **13.** Because ___ are the preferred energy source for the body, they should make up the greatest percentage of the diet.
A. micronutrients
B. proteins
C. lipids
D. carbohydrates

_______________ **14.** A(n) ___ of all grains consumed should be from whole grain sources, such as whole wheat breads and pastas, brown rice, wild rice, wheat berries, bulgur, oats, barley, rye, quinoa, millet, and spelt.
A. eighth
B. fourth
C. third
D. half

_______________ **15.** ___ is a yellow-green fluid produced by the liver and passed into the small intestine where it enables fats to mix with water for digestion.
A. Gastric juice
B. Amylase
C. Saliva
D. Bile

_______________ **16.** ___ is believed to be the most important nutrient.
A. Water
B. Dietary fiber
C. Protein
D. Starch

_______________ **17.** The ___ is the minimum amount of calories the body needs to support life.
A. basal metabolic rate (BMR)
B. body mass index (BMI)
C. recommended dietary allowance (RDA)
D. estimated average requirement (EAR)

_______________ **18.** Foods made from milk that have little to no calcium, such as ___, are not considered dairy foods.
A. yogurt
B. butter
C. buttermilk
D. cheese

_______________ **19.** Saliva moistens and helps compress food into ___.
A. a bolus
B. an ileum
C. chyme
D. bile

_______________ **20.** ___ is a hormone produced by the pancreas that is necessary for regulating blood sugar levels.
A. Lactose
B. Adenosine triphosphate
C. Insulin
D. Gluten

Completion

_______________ **1.** ___ is the involuntary and wavelike contraction of muscles that transports food through the digestive tract.

_______________ **2.** ___ is an enzyme in saliva that breaks down carbohydrates into simple sugars.

_______________ **3.** The ___ is a measure indicating the rate at which an ingested food causes blood sugar levels to rise.

______________________ **4.** A(n) ___ is the maximum daily intake level of a nutrient before the body experiences an adverse effect.

______________________ **5.** ___ is a thick, semifluid mass of partly digested food and gastric juices that originates in the stomach and is passed to the small intestine.

______________________ **6.** A(n) ___ is an energy-providing nutrient made from fatty acids and includes solid fats and oils.

______________________ **7.** ___ is the metabolic reaction by which the body breaks down the complex molecules in nutrients into simpler molecules in order to release useable energy.

______________________ **8.** The recommended daily serving size of fruits is ___ cups daily.

______________________ **9.** Obesity is a major risk factor for developing type ___ diabetes.

______________________ **10.** The body mass index (BMI) is a calculation based on height and ___ to determine body fat composition.

______________________ **11.** A(n) ___ is a fingerlike projection located in the small intestine that facilitates nutrient absorption.

______________________ **12.** A(n) ___ food is a food that has had nutrients added that were originally present in the food but were lost during processing.

______________________ **13.** A(n) ___ is an organic nutrient that is required in small amounts to help regulate body processes.

______________________ **14.** The Food Allergen Labeling and Consumer Protection Act states that food labels must identify the eight foods that account for 90% of all food allergies: ___, eggs, fish, shellfish, tree nuts, peanuts, wheat, and soy.

______________________ **15.** A(n) ___ nutrient is a nutrient the body cannot make in sufficient amounts to meet its needs and must be obtained from food.

Critical Thinking

Use complete sentences to respond to the following.

1. Explain the role of the mouth, stomach, small intestine, and large intestine in the digestive process.

2. Compare the processes of catabolism and anabolism.

3. Identify ways to reduce the risk of hypertension, obesity, cardiovascular disease, diabetes, and cancer.

4. Summarize the six steps in the culinary nutrition recipe-modification process.

Nutrition Fundamentals — Chapter 2 ACTIVITIES

Name ______________________________ Date ______________

Activity 2-1: Essential Nutrients

An essential nutrient is a nutrient the body cannot make in amounts sufficient to meet its needs and must be obtained from food. The body requires six essential nutrients to function properly. Essential nutrients are also classified as either macronutrients or micronutrients depending on the amount needed by the body. In addition, the acceptable macronutrient distribution range (AMDR) identifies a percentage intake range for the three energy-providing macronutrients.

Respond to the following.

1. Identify the six essential nutrients.

2. Identify which essential nutrients are classified as macronutrients.

3. Identify which essential nutrients are classified as micronutrients.

4. Name the macronutrient with an AMDR of 45%–65%.

5. Name the macronutrient with an AMDR of 10%–35%.

6. Name the macronutrient with an AMDR of 20%–35%.

Activity 2-2: Digestive Tract

The digestive tract serves as the passageway for food to move through the body. As food moves through this passageway, nutrients are broken down and used for energy or used to build and maintain cells.

Match the description below with the appropriate letter on the illustration shown.

______________________ **1.** Stomach

______________________ **2.** Esophagus

______________________ **3.** Gallbladder

______________________ **4.** Anus

______________________ **5.** Mouth

______________________ **6.** Pancreas

______________________ **7.** Large intestine

______________________ **8.** Liver

______________________ **9.** Rectum

______________________ **10.** Small intestine

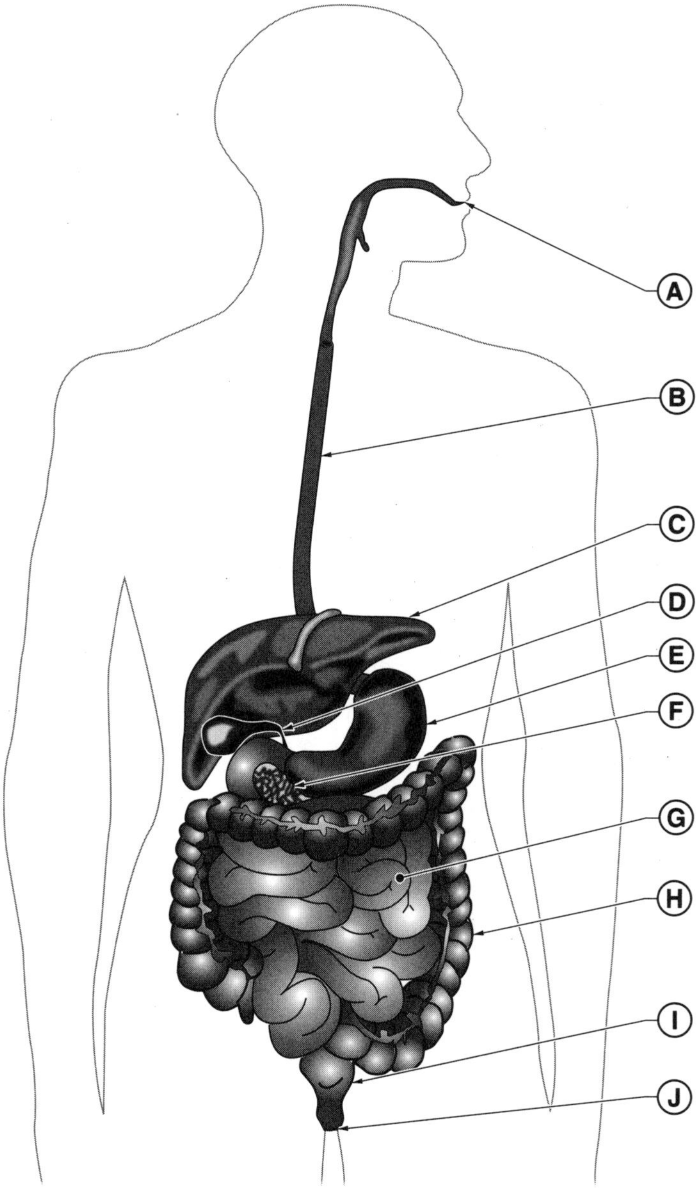

Activity 2-3: Macronutrient Calorie Contributions

Macronutrients such as lipids (fats), carbohydrates, and proteins provide energy and therefore, supply calories. Lipids supply 9 calories of energy per gram, while carbohydrates and proteins each supply 4 calories per gram.

Use the following Nutrition Facts label to calculate the calorie contributions from fat, carbohydrates, and protein.

Nutrition Facts

Serving Size 5 oz
Servings Per Container 4

Amount Per Serving	
Calories 90	
	% Daily Value*
Total Fat 3g	5%
Saturated Fat 0g	0%
Cholesterol 0mg	0%
Sodium 440mg	19%
Total Carbohydrate 13g	4%
Dietary Fiber 3g	4%
Sugars 3g	
Protein 3g	
Vitamin A 80% • Vitamin C	60%
Calcium 4% • Iron	4%

* Percent Daily Values (DV) are based on a 2,000 calorie diet.

________________ **1.** Calories from fat = ___.

________________ **2.** Calories from carbohydrates = ___.

________________ **3.** Calories from protein = ___.

Activity 2-4: Energy (Calorie) Output

The body burns (uses) extra calories through movement and physical activity. When more calories are consumed than are used, weight gain occurs. In order to maintain body weight, the calories consumed (energy intake) and calories used (energy output) must be equal.

Example:

A 154 lb adult burns 140 calories walking at 3½ miles per hour for 30 minutes. In order to burn 500 calories walking at 3½ miles per hour, the individual must walk for 107 minutes.

The length of time it takes the 154 lb adult to burn 500 calories walking at 3½ miles per hour is calculated using the following steps:

1. Divide the number of calories used (140) by the number of minutes walked (30) to calculate the number of calories used per minute.

   140 ÷ 30 = 4.67 calories used per minute

2. Divide the calories that need to be burned (500) by the calories used per minute (4.67) to calculate the length of time the activity must be performed.

   500 ÷ 4.67 = 107 minutes

Calculate the length of time the activities listed in the chart need to be performed in order to burn a total of 500 calories.

	Approximate Calories Used by a 154 lb Adult		
	Physical Activity	**30 Minutes of Activity**	**Length of Time to Burn 500 Calories**
	Walking (3½ miles per hour)	*140 (See example)*	*107 minutes (See example)*
1.	Walking (4½ miles per hour)	230	
2.	Bicycling (less than 10 miles per hour)	145	
3.	Basketball (vigorous)	220	
4.	Swimming (laps)	255	
5.	Running/jogging (5 miles per hour)	295	

Activity 2-5: Serving Size References

A serving size is a specific amount of food expressed using a unit of measurement, such as a tablespoon, an ounce, or a cup. Many dietary guidelines are based on serving size, and a visual reference can be helpful in determining the appropriate amounts of food to consume.

Place an X in the column (A, B, or C) that corresponds to the reference item.

	Reference Item	Serving Size A	Serving Size B	Serving Size C
1.	Computer mouse 	1 small baked potato	1 medium baked potato	1 large baked potato
2.	Baseball 	1 small apple/orange	1 medium apple/orange	1 large apple/orange
3.	Average female fist 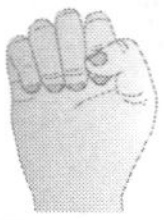	½ cup raw vegetables	¾ cup raw vegetables	1 cup raw vegetables
4.	Tennis ball 	½ cup cooked pasta/grains	¾ cup cooked pasta/grains	1 cup cooked pasta/grains
5.	Deck of cards 	3 oz cooked poultry/meat/seafood	4 oz cooked poultry/meat/seafood	5 oz cooked poultry/meat/seafood
6.	Shot glass 	1 oz nuts	2 oz nuts	3 oz nuts
7.	Die 	1 tsp butter	1½ tsp butter	2 tsp butter
8.	Domino 	1 oz cheese	2 oz cheese	3 oz cheese
9.	Golf ball 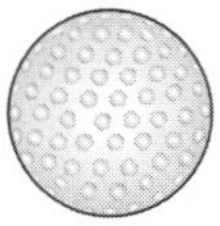	1 tbsp peanut butter	1½ tbsp peanut butter	2 tbsp peanut butter

Activity 2-6: Food Allergens

A food allergy is an adverse reaction to a specific food that involves the immune system. Allergic reactions can be minor or extremely severe. There are over 100 foods that can cause allergic reactions. However, the Food and Drug Administration (FDA) has identified eight specific foods considered major allergens: eggs, soy, milk, wheat, peanuts, tree nuts, fish, and shellfish.

Match each of the following dishes with the allergen(s) it contains.

______ **1.** Banana pudding with cookie crumble

______ **2.** Beef barley soup

______ **3.** Bouillabaisse

______ **4.** Traditional carrot cake

______ **5.** Cheese grits

______ **6.** Classic Caesar salad

______ **7.** Fried egg white sandwich

______ **8.** Greek salad

______ **9.** Hummus with pita

______ **10.** Manhattan clam chowder

______ **11.** New England clam chowder

______ **12.** Orange sherbet

______ **13.** Reuben sandwich with coleslaw

______ **14.** Sole almondine

______ **15.** Steamed edamame

______ **16.** Toasted pine nut couscous

______ **17.** Turkey burger with French fries

______ **18.** Vanilla ice cream

______ **19.** Waldorf chicken salad

______ **20.** Yogurt and granola-cashew parfait

A. Egg, soy, and/or dairy products

B. Wheat (gluten)

C. Peanuts and/or tree nuts

D. Fish and/or shellfish

Activity 2-7: Dietary Analysis

Dietary analysis tools enable individuals to track, monitor, and compare their actual nutrient intake against recommended nutrient values based on factors such as age, gender, and activity level. These tools provide valuable feedback that often leads to positive dietary changes.

Access a dietary analysis tool from one of the following websites to complete this activity.

FatSecret™: www.fatsecret.com
SelfNutritionData: http://nutritiondata.self.com/mynd/mytracking
SparkPeople®: www.sparkpeople.com
SuperTracker: www.supertracker.usda.gov

1. Register your profile information at the website of your choice.
2. Enter all foods and beverages consumed for seven consecutive days, including snacks, and the physical activities performed.
3. Submit a copy of your seven-day dietary analysis and a written report that addresses the following criteria:
   a. Describe how the dietary analysis compares to USDA dietary recommendations in terms of nutrition and physical activity levels.
   b. Identify three changes that could be made to improve nutrient intake.
   c. Identify three changes that could be made to improve physical activity levels.

Dietary Analysis Rubric

Content	
Copy of dietary analysis	10
Written report	5
Quality content	10
Summary	5
Format	
Typed, double spaced	5
Correct spelling	5
Correct grammar	5
Correct punctuation	5
Total	50

Activity 2-8: Nutrition and Physical Activity Myths Research

There are many nutrition and physical activity myths that are believed to be true. Understanding what is myth and what is factual provides individuals with knowledge that can lead to optimal health and well-being.

Access the Weight-Control Information Network website at http://win.niddk.nih.gov/publications/myths.htm. Prepare a three- to four-page report and an oral presentation that address the following criteria.

1. Identify five myths pertaining to nutrition and physical activity and explain why the information is not factual.
2. Give an example of a breakfast, a lunch, and a dinner that incorporate complex carbohydrates.
3. Identify three fat-free and low-fat foods that are high in calories and explain why they are high in calories.
4. Explain why skipping meals is not advisable.
5. Give an example of a healthy fast food meal and explain why it is healthy.
6. Describe three nutrition practices that promote health.
7. Explain three ways physical activity can be incorporated throughout the day.

Nutrition Project Rubric

Content	
Written report	5
Cover page	5
Introduction	5
Quality content	20
Summary	5
Format	
Typed, double spaced	5
Correct spelling	5
Correct grammar	5
Correct punctuation	5
Bibliography	5
Oral Presentation	
Introduction	5
Quality content	10
Summary	5
Delivery	5
Visual aid(s)	10
Total	100

Proteins Chapter 3 REVIEW

Name ______________________________ Date ____________________

True-False

T F **1.** Protein foods should be prepared using high heat to reduce shrinkage.

T F **2.** Acids cause denaturation to occur in protein foods.

T F **3.** The most important functions of proteins are to build, repair, and maintain the cells and tissues of the body.

T F **4.** Denatured proteins lose their biological activity as enzymes but not their nutritional value.

T F **5.** If the pH of blood varies from an optimal level of 7.4, proteins neutralize the acids and bases by releasing hydrogen ions when the blood is too basic.

T F **6.** The majority of plant-based foods are complete proteins.

T F **7.** Protein turnover is the continuous breakdown, rebuilding, and recycling of amino acids.

T F **8.** The backbone of an amino acid can vary.

T F **9.** Myoglobin is a soft, white connective tissue that breaks down into gelatin when heated.

T F **10.** Medical research indicates that a diet containing excessive amounts of protein has no adverse health effects.

T F **11.** A smaller portion of protein can be paired with additional vegetables and healthy grains to offer a more nutritious version of a meal.

T F **12.** In liquids such as milk, acids can cause proteins to curdle.

T F **13.** A nonessential amino acid is an amino acid that can be made by the body in sufficient quantities.

T F **14.** Protein needs cannot be met by consuming a variety of incomplete proteins throughout the day.

T F **15.** Thinly slicing and fanning cooked meats, poultry, or seafood gives the appearance of a larger portion size.

Multiple Choice

______ **1.** A(n) ___ amino acid is an amino acid the body cannot make in sufficient quantities and must be obtained from dietary sources.
A. essential
B. nonessential
C. complementary
D. complete

______ **2.** If the body lacks a sufficient supply of carbohydrates and lipids for energy, proteins will sacrifice their vital functions to provide ___ calories of energy per gram.
A. 2
B. 4
C. 6
D. 8

______ **3.** During protein absorption, the ___ determines where the amino acids are needed and sends them into the bloodstream to the required cells.
A. thyroid gland
B. gallbladder
C. pancreas
D. liver

______ **4.** The ___ structure of a protein is a coiled chain that has the appearance of a spring.
A. primary
B. secondary
C. tertiary
D. quaternary

______ **5.** ___ is the element that differentiates proteins from carbohydrates and lipids.
A. Carbon
B. Hydrogen
C. Nitrogen
D. Oxygen

______ **6.** ___ and quinoa contain complete proteins.
A. Soybeans
B. Barley
C. Wheat
D. Oats

______ **7.** When proteins are subjected to ___, they begin to coagulate.
A. air
B. heat
C. moisture
D. alkalis

_______________ **8.** ___ is a protein in red blood cells that transports oxygen throughout the body.
A. Pepsin
B. Fibrin
C. Myoglobin
D. Hemoglobin

_______________ **9.** ___ is a form of protein energy malnutrition characterized by a diet deficient in both proteins and caloric intake.
A. Gout
B. Ketosis
C. Marasmus
D. Kwashiorkor

_______________ **10.** To determine individual protein needs more accurately, ___ g of protein should be consumed per pound of body weight.
A. 0.12
B. 0.36
C. 0.98
D. 1.23

Completion

_______________ **1.** ___ are commonly referred to as "the building blocks of protein."

_______________ **2.** A(n) ___ is a type of protein that acts as a catalyst to accelerate chemical reactions.

_______________ **3.** A(n) ___ protein is a combination of two or more incomplete proteins that supplies sufficient amounts of all the essential amino acids.

_______________ **4.** A(n) ___ is a bond that connects a nitrogen-containing amino group of one amino acid to the acid group of another amino acid to form a link in a protein chain.

_______________ **5.** ___ is a digestive enzyme in the small intestine that breaks the peptide bonds connecting amino acids.

_______________ **6.** ___ is an iron-containing protein that binds oxygen in muscle tissues.

_______________ **7.** A(n) ___ is a type of protein produced by the immune system that destroys or deactivates antigens.

_______________ **8.** ___ is the process of structurally changing protein from its natural state.

_______________ **9.** A(n) ___ amino acid is an amino acid that is missing or unavailable in an adequate amount.

_______________ **10.** The ___ reaction is a reaction that occurs when the proteins and sugars in a food are exposed to heat and merge together to form a brown exterior.

Critical Thinking

Use complete sentences to respond to the following.

1. Explain how amino acids form proteins.

2. Describe how protein excess and protein deficiencies can affect the body.

3. Describe the reactions of proteins to air, heat, moisture, and acids.

Proteins | Chapter 3 | ACTIVITIES

Name ______________________________ Date ____________________

Activity 3-1: Proteins

The word "protein" is derived from the Greek word "proteos," which means "prime importance." Proteins are of prime importance because they are involved in a variety of functions that support life. The amino acid sequence enables each protein to perform a specific function. Some proteins work alone, while other proteins work together to carry out functions that are vital to health and well-being.

Complete the following.

1. Identify the four elements that comprise a protein.

2. Name the element that differentiates proteins from carbohydrates and lipids.

Indicate which of the following are the nine essential functions of proteins by responding "yes" or "no" to each.

______________ **3.** Growth and maintenance

______________ **4.** Insulation and organ protection

______________ **5.** Transport substances

______________ **6.** Energy

______________ **7.** Promote elimination of waste

______________ **8.** Antibody production

______________ **9.** Hormone production

______________ **10.** Nutrient absorption

______________ **11.** Enzyme production

______________ **12.** Cellular structure

______________ **13.** Blood clotting

______________ **14.** Influence communication among cells

______________ **15.** Acid-base balance

______________ **16.** Supply raw materials to build compounds

______________ **17.** Fluid balance

Activity 3-2: Protein Digestion and Absorption

During digestion, proteins are broken down into amino acids. This process begins in the stomach and continues in the small intestine where the majority of nutrient absorption occurs.

Match the description below with the appropriate letter on the illustration shown.

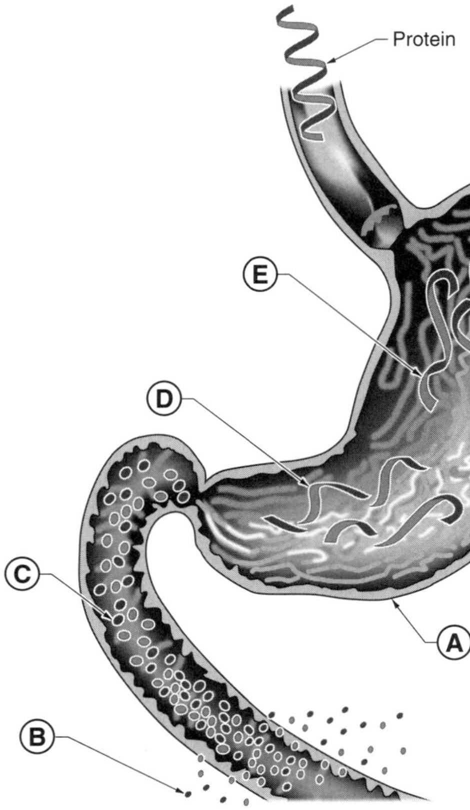

_______________ **1.** Amino acids enter small intestine where protease enzymes continue to break peptide bonds

_______________ **2.** Protein enters stomach and is denatured by stomach acid

_______________ **3.** Stomach

_______________ **4.** Amino acids are absorbed by small intestine and sent to bloodstream

_______________ **5.** Pepsin enzymes break peptide bonds and shorten amino acid chains

Activity 3-3: Protein Dietary Analysis

Proteins build, repair, and maintain all cells and tissues throughout the body and support vital internal functions. Excess dietary protein can lead to conditions such as dehydration, gout, osteoporosis, and ketosis. In contrast, protein deficiency causes the body to break down muscles to use as an energy source, causing wasting of muscles, organs, and tissues.

Access your online dietary analysis profile that was created in Activity 2-7 from one of the following websites:

FatSecret™: www.fatsecret.com
SelfNutritionData: http://nutritiondata.self.com/mynd/mytracking
SparkPeople®: www.sparkpeople.com
SuperTracker: www.supertracker.usda.gov

1. Submit a copy of your seven-day dietary analysis.
2. Submit a written report that includes the following criteria:
   a. Describe how the daily protein intake compares with USDA dietary recommendations for your age, gender, and physical activity level.
   b. Compare the amount of protein consumed from animal-based sources with the amount consumed from plant-based sources.
   c. Identify three changes that could be made to improve protein consumption.

Dietary Analysis Rubric

Content	
Copy of dietary analysis	10
Written report	5
Quality content	10
Summary	5
Format	
Typed, double spaced	5
Correct spelling	5
Correct grammar	5
Correct punctuation	5
Total	50

Activity 3-4: Cooking Techniques and the Maillard Reaction Lab

The Maillard reaction occurs when the proteins and sugars in a food are exposed to heat and merge together to form a brown exterior. As food browns, aromatic compounds are released and flavor is developed. Some cooking techniques facilitate the Maillard reaction, while others do not.

Objective: To determine the cooking techniques that cause the Maillard reaction and promote flavor development.

Materials and Equipment

- 1 lb lean ground beef, pork, chicken, or turkey
- Sauté pan with a lid
- Nonstick cooking spray
- 1 tbsp water
- Vegetable oil
- Instant read thermometer
- Spatula
- Sheet pan
- Broiler pan
- Tasting forks

Note: Beef and pork patties should be cooked to an internal temperature of 160°F, and chicken and turkey patties should be cooked to an internal temperature of 165°F.

Procedure

1. Divide the meat into four equal portions and shape the portions into patties.
2. Place one patty in a heated sauté pan that has been sprayed with nonstick cooking spray.
3. Add 1 tbsp of water and cover the pan with a lid to capture the steam. Cook to the appropriate degree of doneness. Evaluate the patty in terms of appearance and flavor.
4. Place the second patty on a heated grill that has been coated with oil. Grill on one side. Flip the patty and cook to the appropriate degree of doneness. Evaluate the patty in terms of appearance and flavor.
5. Place the third patty on a sheet pan that has been sprayed with nonstick cooking spray. Bake at 350°F to the appropriate degree of doneness. Evaluate the patty in terms of appearance and flavor.
6. Place the fourth patty on a broiler pan that has been sprayed with nonstick cooking spray. Broil 3–4 inches from the heat source to the appropriate degree of doneness. Evaluate the patty in terms of appearance and flavor.

Respond to the following questions.

1. Which cooking techniques caused the Maillard reaction?

2. Which patties were the most appealing in terms of appearance and flavor?

Activity 3-5: Protein Supplements Research

Protein supplements are often used by athletes and are available in bars, powders, and beverages. These items typically include protein that has been derived from soy, egg whites, or milk in the form of whey and casein.

Access an article from a medical-based website, a nutrition journal, and a company that sells protein supplements. Prepare a three- to four-page report and an oral presentation that address the following criteria.

1. Describe three advantages of protein supplements cited in the articles.
2. Describe three disadvantages of protein supplements cited in the articles.
3. Explain when protein supplements might be appropriate for certain athletes to use.
4. Identify three medical reasons for using protein supplements.
5. Describe the long-term repercussions associated with protein supplements.
6. Identify three ways to increase protein consumption without the use of protein supplements.
7. Summarize whether or not protein supplements are beneficial.

Nutrition Project Rubric

Content	
Written report	5
Cover page	5
Introduction	5
Quality content	20
Summary	5
Format	
Typed, double spaced	5
Correct spelling	5
Correct grammar	5
Correct punctuation	5
Bibliography	5
Oral Presentation	
Introduction	5
Quality content	10
Summary	5
Delivery	5
Visual aid(s)	10
Total	100

Carbohydrates Chapter 4 REVIEW

Name ______________________________ **Date** ______________

True-False

T F **1.** Sugars brown as a result of the Maillard reaction.

T F **2.** Dehydration is the process of adding moisture to food.

T F **3.** Gelatinization occurs as heated starch granules absorb moisture and swell.

T F **4.** According to the Institute of Medicine, a healthy diet should consist of approximately 45%–65% carbohydrates.

T F **5.** When dietary fiber is added to the diet on a consistent basis, the risk of contracting cardiovascular disease is found to decrease.

T F **6.** Whole grains are easier to digest, have a longer shelf life, and take less time to cook than refined grains.

T F **7.** In baking, fermentation is accomplished by using yeast to convert proteins into carbon dioxide and alcohol.

T F **8.** Artificial sweeteners are considered nonnutritive.

T F **9.** Complex carbohydrates are mainly found in vegetables, fruits, legumes, and whole grains.

T F **10.** Carbohydrates need to be broken down into polysaccharides in order to be absorbed through the cells that line the small intestine.

T F **11.** The brain and central nervous system cannot use lipids for energy.

T F **12.** Phytonutrients are naturally occurring substances that increase the risk of disease.

T F **13.** Carbohydrates are made from the elements carbon, hydrogen, and oxygen.

T F **14.** Soluble fiber does not dissolve in water.

T F **15.** Simple carbohydrates provide sustained energy and promote satiety.

Multiple Choice

_______________ **1.** Carbohydrates contribute ___ calories of energy per gram.
- A. 2
- B. 4
- C. 6
- D. 8

_______________ **2.** ___ are considered the simplest forms of sugar and are sometimes called the "building blocks of carbohydrates."
- A. Oligosaccharides
- B. Disaccharides
- C. Monosaccharides
- D. Polysaccharides

_______________ **3.** ___ is a polysaccharide made and used by the liver and muscles to store long chains of glucose.
- A. Pectin
- B. Lignin
- C. Insulin
- D. Glycogen

_______________ **4.** A whole grain is a grain that has only had the ___ removed.
- A. husk
- B. bran
- C. germ
- D. endosperm

_______________ **5.** A saccharide is a unit of ___.
- A. sugar
- B. protein
- C. fat
- D. amino acid

_______________ **6.** The term "___" is the only term that ensures a product contains a whole grain.
- A. wheat
- B. wheat flour
- C. unbleached flour
- D. whole grain

_______________ **7.** ___ is an insoluble component of starch that forms an unbranched chain.
- A. Amylose
- B. Amylopectin
- C. Pectin
- D. Glucagon

________________ **8.** At least half of all grains consumed should be ___ grains.
A. refined
B. whole
C. pearled
D. milled

________________ **9.** Photosynthesis is the process by which plants turn ___ and water into carbohydrates by using chlorophyll to capture energy from sunlight.
A. oxygen
B. hydrogen
C. carbon dioxide
D. carbon monoxide

________________ **10.** ___ is often mistakenly considered an artificial sweetener because it is intensely sweet and has no calories.
A. Aspartame
B. Stevia
C. Agave nectar
D. High-fructose corn syrup

________________ **11.** The ___ is the largest component of a grain kernel and consists of carbohydrates and a small amount of protein.
A. husk
B. bran
C. germ
D. endosperm

________________ **12.** The glycemic index (GI) is a measure indicating the rate at which an ingested food ___.
A. passes through the digestive tract
B. supplies essential nutrients
C. causes blood sugar levels to rise
D. causes blood pressure to rise

________________ **13.** ___, also known as milk sugar, is a disaccharide comprised of glucose and galactose.
A. Sucrose
B. Lactose
C. Fructose
D. Maltose

________________ **14.** All digestible forms of simple and complex carbohydrates must be converted to ___ for energy use.
A. glucose
B. sucrose
C. starch
D. fiber

__________ **15.** ___ is a condition that results from low blood glucose levels.
- A. Hypoglycemia
- B. Hyperglycemia
- C. Cardiovascular disease
- D. Dehydration

__________ **16.** Most adults require a minimum of ___ g of carbohydrates daily.
- A. 100
- B. 110
- C. 120
- D. 130

__________ **17.** Polysaccharides are classified as ___.
- A. sucrose and maltose
- B. glucose and fructose
- C. starches and dietary fiber
- D. monosaccharides and disaccharides

__________ **18.** Sugar replacers are typically less sweet than sucrose and provide ___ calories of energy per gram.
- A. 1 to 2
- B. 2 to 3
- C. 3 to 4
- D. 4 to 5

__________ **19.** Glycogen is primarily stored in the liver and ___.
- A. brain
- B. pancreas
- C. intestines
- D. muscles

__________ **20.** ___ fiber is dietary fiber that dissolves in water.
- A. Simple
- B. Complex
- C. Soluble
- D. Insoluble

__________ **21.** When acidic ingredients are used in pie fillings, barbecue sauces, or glazes, ___ are often used to help maintain the desired viscosity.
- A. modified starches
- B. invert sugars
- C. flavonoids
- D. carotenoids

__________ **22.** A carotenoid is an organic pigment found in ___ vegetables.
- A. white
- B. green
- C. orange and yellow
- D. purple and red

____________________ **23.** A ___ is a seed that is classified as a grain but is derived from broadleaf plants instead of grasses.
A. cereal grain
B. pseudocereal
C. refined grain
D. whole grain

____________________ **24.** ___ is a process in which crystals precipitate from a solution to form a solid in the presence of heat.
A. Gelatinization
B. Retrogradation
C. Caramelization
D. Crystallization

____________________ **25.** ___ foods have been found to help with weight management and may reduce the risk of cardiovascular disease, colon cancer, and digestive disorders such as diverticulosis.
A. Sugar-based
B. Fiber-rich
C. Processed
D. Refined

Completion

____________________ **1.** ___ is the breakdown of carbohydrates due to the actions of bacteria, microorganisms, or yeast.

____________________ **2.** A(n) ___ is an acidic substance that forms in the blood when lipids are metabolized for energy instead of carbohydrates.

____________________ **3.** Insulin is a hormone produced by the ___ that is necessary for regulating blood sugar levels.

____________________ **4.** A(n) ___ carbohydrate is a carbohydrate made from more than two attached sugar units.

____________________ **5.** ___ is a process in which a gelatinized starch causes a thickened liquid to turn into a gel as it cools.

____________________ **6.** ___ is an indigestible form of carbohydrate found in vegetables, fruits, and grains.

____________________ **7.** A(n) ___ is any sugar not naturally occurring in a food and added during processing or preparation.

____________________ **8.** ___ is a soluble component of starch that forms a branched chain.

____________________ **9.** ___ is a pancreatic hormone sent to the liver to trigger the conversion of glycogen to glucose for energy.

____________________ **10.** ___ is the process of splitting a substance into smaller parts by the addition of water.

Critical Thinking

Use complete sentences to respond to the following.

1. Describe the various forms of simple carbohydrates.

2. Describe the various forms of complex carbohydrates.

3. Explain how glucose is used and stored for energy.

Name ______________________________ Date ________________

Activity 4-1: Carbohydrates

A carbohydrate is an energy-providing nutrient in the form of sugar, starch, or dietary fiber. Carbohydrates are the main source of energy for the body. They are comprised of sugar units known as saccharides. Simple carbohydrates contain fewer saccharides (sugar units) than complex carbohydrates.

A monosaccharide is a simple carbohydrate made of one sugar unit. There are three types of monosaccharides: glucose, fructose, and galactose. Before carbohydrates can be used for energy, they must be converted into their simplest form, which is glucose.

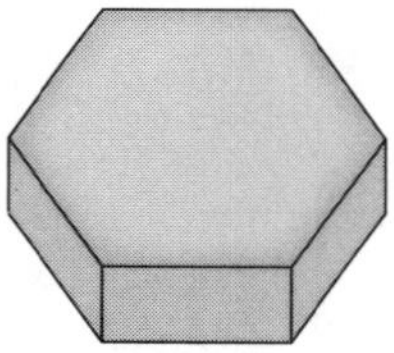

Glucose

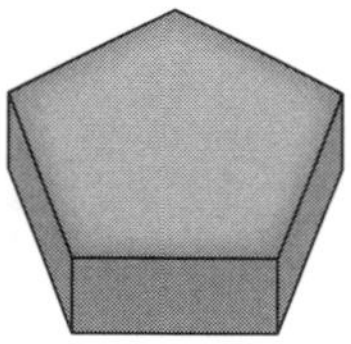

Fructose

Galactose

A disaccharide is a simple carbohydrate made of two attached sugar units. Disaccharides contain glucose in combination with itself or in combination with fructose or galactose.

Identify the monosaccharides that combine to form the following disaccharides.

1. Sucrose = ____________________ + ____________________

2. Lactose = ____________________ + ____________________

3. Maltose = ____________________ + ____________________

A polysaccharide is a complex carbohydrate with a chain usually consisting of hundreds to thousands of saccharides. Polysaccharides are classified as starches and dietary fiber. Amylose and amylopectin are polysaccharides in the form of starch.

Match each of the following characteristics with the appropriate polysaccharide.

_______________ **4.** Opaque when cooked

_______________ **5.** Soft gel

_______________ **6.** Thickens with cooling

_______________ **7.** Freezes and thaws well

_______________ **8.** Transparent when cooked

_______________ **9.** Separates when frozen

_______________ **10.** Thins with cooling

_______________ **11.** Firm gel

A. Amylose

B. Amylopectin

Activity 4-2: Carbohydrate Digestion and Absorption

Carbohydrates need to be broken down into monosaccharides in order to be absorbed by the small intestine. The liver converts monosaccharides to glucose, which is then sent to the cells to be used for energy, stored by the liver and muscles as glycogen, or converted to fat.

Match the description below with the appropriate letter on the illustration shown.

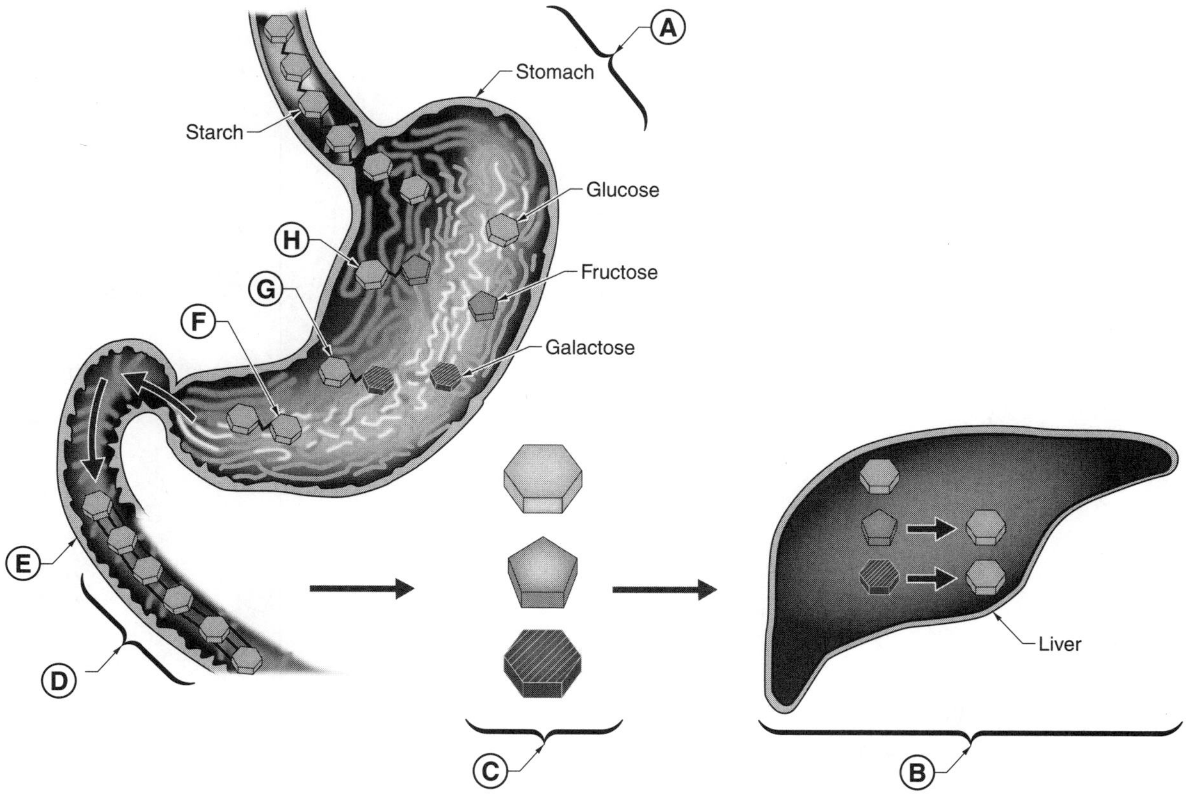

________________ **1.** Carbohydrates are broken down into glucose, fructose, and galactose and sent to liver

________________ **2.** Lactose

________________ **3.** Digestion of carbohydrates comes to a halt in the stomach acids

________________ **4.** Dietary fibers travel to large intestine

________________ **5.** Liver converts fructose and galactose to glucose for energy

________________ **6.** Maltose

________________ **7.** Sucrose

________________ **8.** Small intestine

Activity 4-3: Carbohydrate Dietary Analysis

Carbohydrates are the main source of energy for the body. Consumption of simple carbohydrates, or foods with added sugar and empty calories, may put the body at greater risk for obesity, diabetes, cardiovascular disease, and tooth decay. In contrast, diets plentiful in complex carbohydrates that include dietary fiber can reduce these risks. Carbohydrate deficiency can cause a decrease in energy, leading to dizziness, confusion, digestive problems, and weight loss.

Access your online dietary analysis profile that was created in Activity 2-7 from one of the following websites:

FatSecret™: www.fatsecret.com
SelfNutritionData: http://nutritiondata.self.com/mynd/mytracking
SparkPeople®: www.sparkpeople.com
SuperTracker: www.supertracker.usda.gov

1. Submit a copy of your seven-day dietary analysis.
2. Submit a written report that includes the following criteria:
   a. Describe how the daily carbohydrate intake compares with USDA dietary recommendations for your age, gender, and physical activity level.
   b. Compare the amount of simple carbohydrates consumed with the amount consumed from complex carbohydrates.
   c. Identify three changes that could be made to improve carbohydrate consumption.

Dietary Analysis Rubric

Content	
Copy of dietary analysis	10
Written report	5
Quality content	10
Summary	5
Format	
Typed, double spaced	5
Correct spelling	5
Correct grammar	5
Correct punctuation	5
Total	50

Activity 4-4: Sugars and Sugar Substitutes Lab

Sugar comes in a variety of forms, each with a different appearance and flavor. These various sugars are often added to menu items to enhance colors, textures, and flavors. Sugar substitutes are generally calorie-free and sweeter than most forms of sugar.

Objective: To identify sugars and sugar substitutes based on appearance and flavor.

Materials and Equipment

- Marker
- 15 tasting cups
- Teaspoon
- Sugars
  - agave nectar
  - confectioners' sugar
  - corn syrup
  - dark brown sugar
  - granulated white sugar
  - honey
  - light brown sugar
  - maple syrup
  - molasses
  - turbinado sugar
- Sugar substitutes
  - aspartame
  - acesulfame K
  - saccharin
  - sucralose
  - stevia
- 30 tasting spoons

Allergen Alert

The materials used in this activity may contain one or more food allergens.

Procedure

Sugars

1. Label the bottoms of ten of the tasting cups, one type of sugar per label.
2. Place 1 tsp of each type of sugar into the correctly labeled cup.
3. Working in pairs, take turns placing the cups of sugars in random order and identifying each sugar based on its appearance.
4. Place a blindfold on one lab partner and have that partner taste and identify each sugar based on its flavor.
5. Repeat step 4 with the other lab partner wearing the blindfold.

Sugar substitutes

6. Label the bottoms of the five remaining tasting cups, one type of sugar substitute per label.
7. Place 1 tsp of each sugar substitute into the correctly labeled cup.
8. Working in pairs, take turns placing the cups of sugar substitutes in random order and identifying each sugar substitute based on its appearance.
9. Place a blindfold on one lab partner and have that partner taste and identify each sugar substitute based on its flavor.
10. Repeat step 9 with the other lab partner wearing the blindfold.

Scoring: A point is earned for each sugar and sugar substitute correctly identified based on its appearance and flavor for a possible total score of 30 points.

Activity 4-5: Diabetes Research

Diabetes is often treated with oral medications or insulin. In addition, diabetics are advised to follow a diet that promotes safe blood glucose levels. This entails monitoring the amount of carbohydrates consumed based on factors such as age, activity level, and overall health.

Access one of the following websites:

American Diabetes Association: www.diabetes.org/
Center for Disease Control and Prevention: www.cdc.gov/diabetes
MedlinePlus: www.nlm.nih.gov/medlineplus/diabetes.html
National Diabetes Information Clearinghouse: www.diabetes.niddk.nih.gov

Prepare a three- to four-page report and oral presentation that address the following criteria.

1. Differentiate between type 1 and type 2 diabetes.
2. Explain how diabetes can affect the body if it is left untreated.
3. Identify risk factors for developing diabetes.
4. Describe dietary recommendations for diabetics.
5. Describe a breakfast, a lunch, and a dinner (including beverages) that meet the dietary recommendations for diabetics.

Nutrition Project Rubric

Content	
Written report	5
Cover page	5
Introduction	5
Quality content	20
Summary	5
Format	
Typed, double spaced	5
Correct spelling	5
Correct grammar	5
Correct punctuation	5
Bibliography	5
Oral Presentation	
Introduction	5
Quality content	10
Summary	5
Delivery	5
Visual aid(s)	10
Total	100

Lipids Chapter 5 REVIEW

Name ______________________________ **Date** ____________________

True-False

T F **1.** Foods that contain trans fats should be eaten in moderation.

T F **2.** Dietary cholesterol is only found in animal-based foods.

T F **3.** Many experts recommend limiting saturated fats to less than 10% of the total calories consumed in a day.

T F **4.** Once lipids reach the stomach, they are exposed to bile.

T F **5.** Typically, the longer the fatty acid chain, the more solid the lipid will be at room temperature.

T F **6.** Lipids form a layer of "fat" under the skin that insulates the body.

T F **7.** Products containing less than 0.5 g of trans fat per serving can be labeled as having 0 g of trans fat.

T F **8.** An effective way to incorporate unsaturated fats into a menu is to replace oils with solid fats.

T F **9.** Ergosterol is a sterol that is present in animal-based foods.

T F **10.** Due to their structural composition, unsaturated fats are more sensitive to air.

T F **11.** The essential fatty acids are found primarily in meats and poultry.

T F **12.** Nuts and seeds are beneficial sources of unsaturated fats.

T F **13.** Fats and oils lubricate the taste buds and allow flavors to be dispersed throughout the palate.

T F **14.** A diet lacking in fat has no negative effects on health.

T F **15.** The smoke point of a fat or oil is determined by the temperature that causes the fat or oil to break down, smoke, and emit an odor.

Multiple Choice

_______________ **1.** ___ fats are rare in natural foods but abundant in processed foods due to hydrogenation.
- A. Saturated
- B. Polyunsaturated
- C. Monounsaturated
- D. Trans

_______________ **2.** A ___-density lipoprotein is a type of lipoprotein used to collect cholesterol and bring it to the liver so that it can be dismantled and recycled in bile or excreted.
- A. very-low
- B. low
- C. high
- D. very-high

_______________ **3.** A(n) ___ fatty acid is a fatty acid chain in which each carbon atom is filled to capacity with hydrogen atoms.
- A. essential
- B. nonessential
- C. saturated
- D. unsaturated

_______________ **4.** Lipids supply ___ calories of energy per gram compared to 4 calories per gram from proteins and carbohydrates.
- A. 5
- B. 7
- C. 9
- D. 11

_______________ **5.** ___ fats contain the two fatty acids that are considered essential fatty acids: linoleic acid and alpha-linolenic acid.
- A. Trans
- B. Saturated
- C. Monounsaturated
- D. Polyunsaturated

_______________ **6.** ___ make up 95% of all lipids and are found in the body as well as in food.
- A. Sterols
- B. Phospholipids
- C. Triglycerides
- D. Monoglycerides

_______________ **7.** Linoleic acid is commonly referred to as ___.
- A. omega-6 fatty acid
- B. omega-3 fatty acid
- C. monounsaturated fat
- D. nonessential fatty acid

______ **8.** Many health organizations recommend that fat intake for healthy adults fall within the range of ___ of total caloric intake.
A. 5% to 10%
B. 15% to 20%
C. 20% to 35%
D. 30% to 45%

______ **9.** The structure of a phospholipid is similar to the structure of a triglyceride, except a phospholipid contains only ___ fatty acid chains.
A. two
B. three
C. four
D. five

______ **10.** A diet too low in fat can interfere with the absorption of the fat-soluble vitamins ___.
A. A, B, C, and D
B. A, C, D, and E
C. A, C, D, and K
D. A, D, E, and K

Completion

______ **1.** ___ fat is fat that is stored within the abdomen and surrounds the internal organs.

______ **2.** An unsaturated fatty acid is a fatty acid chain that lacks some ___ atoms.

______ **3.** A(n) ___ is a protein-coated lipid that acts as an emulsifier for larger lipid fragments and cholesterol to facilitate their transport into the lymph and bloodstream for absorption.

______ **4.** The ___ is the point on a fatty acid chain where a missing hydrogen atom is replaced with a double carbon bond.

______ **5.** ___ is a sterol in the lipid family that is a soft, waxy substance made by the body and found in every cell and in foods of animal origin.

______ **6.** ___, commonly referred to as omega-3 fatty acid, is an essential polyunsaturated fatty acid.

______ **7.** ___ lipoproteins contain a higher ratio of lipids than protein and are commonly referred to as bad cholesterol because they are most responsible for depositing cholesterol into the arteries.

______ **8.** A(n) ___ cell is a cell with the primary function of storing excess fat for later energy use.

______ **9.** ___ is the chemical process of using heat to force hydrogen atoms into an unsaturated fatty acid to make it similar in structure to a saturated fat.

______ **10.** A(n) ___ is a substance that mixes with two unlike liquids to produce an emulsion.

Critical Thinking

Use complete sentences to respond to the following.

1. Describe the structure of a triglyceride and include a comparison of saturated, unsaturated, and trans fatty acid chains.

2. Compare the functions of chylomicrons, very-low-density lipoproteins, low-density lipoproteins, and high-density lipoproteins.

3. Describe the reactions of lipids to air, light, heat, and moisture.

Lipids **Chapter 5** ACTIVITIES

Name ____________________ **Date** ____________

Activity 5-1: Lipids

Lipids are the most calorie-dense of the macronutrients and are necessary for optimal health. Lipids are categorized as triglycerides, phospholipids, and sterols. Triglycerides can be further categorized as saturated fat, unsaturated fat, and trans-fat. Due to their structure, phospholipids serve as emulsifiers. Two types of sterols that help the body function properly include cholesterol and ergosterol.

Match each of the following characteristics with the appropriate type of fat.

A. Saturated fat
B. Unsaturated fat
C. Trans fat

__________ **1.** Commonly found in plant-based foods

__________ **2.** Fatty acid chain is filled with hydrogen atoms

__________ **3.** Naturally solid at room temperature

__________ **4.** Rarely found in natural foods

__________ **5.** Naturally has a higher smoke point

__________ **6.** Created to have a longer shelf life

__________ **7.** Fatty acid chain is naturally straight

__________ **8.** Commonly found in animal-based foods

__________ **9.** Liquid at room temperature

__________ **10.** Hydrogen atoms are on opposite sides of a double carbon bond

__________ **11.** Fatty acid chain is typically bent

__________ **12.** Fatty acid chain lacks hydrogen atoms

__________ **13.** Naturally has as a lower smoke point

__________ **14.** Made by the chemical process of hydrogenation

Respond to the following.

15. Contrast the structure of a phospholipid with that of a triglyceride.

16. Explain how a phospholipid acts as an emulsifier.

Match each of the following characteristics with the appropriate type of sterol.

____________________ **17.** Found in mushrooms and yeast

____________________ **18.** Naturally made by the body

____________________ **19.** Plant-based

____________________ **20.** Source of vitamin D for vegetarians

____________________ **21.** Animal-based

A. Cholesterol

B. Ergosterol

Activity 5-2: Lipid Digestion and Absorption

Match the description below with the appropriate letter on the illustration shown.

_______________ **1.** Gallbladder

_______________ **2.** Pancreatic lipase breaks lipids into glycerol, fatty acids, and monoglycerides

_______________ **3.** Lipid

_______________ **4.** Bile

_______________ **5.** Emulsified lipid

_______________ **6.** Stomach

_______________ **7.** Small intestine

_______________ **8.** Lipids float atop water-based stomach fluids

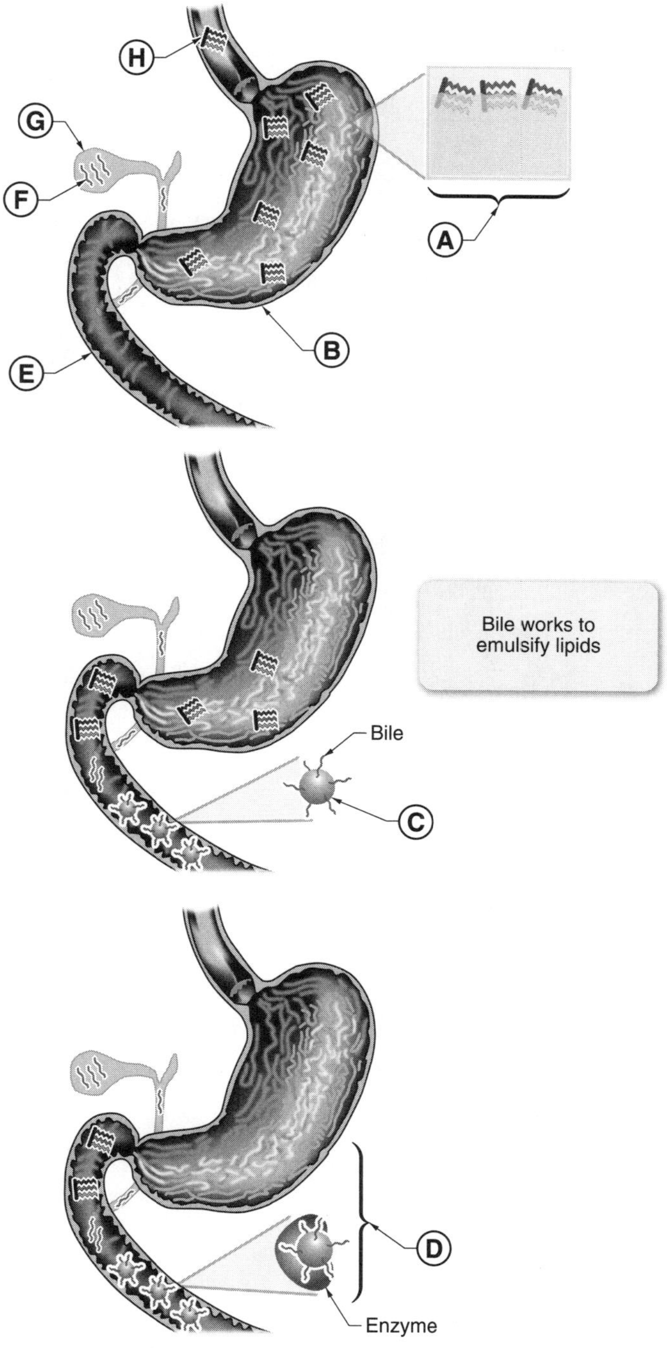

Activity 5-3: Saturated and Unsaturated Fats

Saturated fat has been found to raise the level of LDL, or "bad" cholesterol, which puts the body at risk for developing cardiovascular disease. For this reason, it is recommended that saturated fat consumption be limited and that more emphasis be placed on consuming heart-healthy unsaturated fats.

Identify each of the following foods as containing either saturated fat or unsaturated fat.

____________________ **1.** Butter

____________________ **2.** Olive oil

____________________ **3.** Avocado

____________________ **4.** Nuts

____________________ **5.** Bacon

____________________ **6.** Canola oil

____________________ **7.** Egg yolks

____________________ **8.** Coconut oil

Activity 5-6: Cardiovascular Disease Research

Cardiovascular disease, also known as heart disease, is a class of medical conditions that affect the heart and blood vessels.

Access the American Heart Association website at www.heart.org. Prepare a three- to four-page report and an oral presentation that address the following criteria.

1. Describe the role diet plays in the development of cardiovascular disease.
2. Differentiate between HDL and LDL cholesterol.
3. Explain the role of dietary fiber in reducing the risk of cardiovascular disease.
4. Explain why physical activity can help prevent cardiovascular disease.

Nutrition Project Rubric

Content	
Written report	5
Cover page	5
Introduction	5
Quality content	20
Summary	5
Format	
Typed, double spaced	5
Correct spelling	5
Correct grammar	5
Correct punctuation	5
Bibliography	5
Oral Presentation	
Introduction	5
Quality content	10
Summary	5
Delivery	5
Visual aid(s)	10
Total	100

Water, Vitamins & Minerals

Chapter 6 REVIEW

Name ______________________________ **Date** ____________________

True-False

T F **1.** The heat capacity of water helps the body maintain a consistent body temperature.

T F **2.** Minerals are lost when foods are exposed to air.

T F **3.** Fat-soluble vitamins include vitamins A, C, E, and K.

T F **4.** Bacteria in the intestines can synthesize some vitamin K.

T F **5.** The body does not store water.

T F **6.** Water-soluble vitamins are stored by the body.

T F **7.** Pellegra is a disease caused by a niacin deficiency.

T F **8.** The richest sources of potassium are fresh fruits and vegetables.

T F **9.** Cooked vegetables that are still vibrant in color are more abundant in nutrients than overcooked vegetables with muted colors.

T F **10.** Vitamin E strengthens the immune system and widens the blood vessels.

T F **11.** Folate is more easily absorbed by the body than folic acid.

T F **12.** Vitamin C prevents the absorption of iron from foods.

T F **13.** Green leafy vegetables are significant sources of magnesium.

T F **14.** B vitamins include thiamin, riboflavin, niacin, pantothenic acid, vitamin B_6, biotin, folate, and zinc.

T F **15.** Iodine is required to make hormones in the thyroid gland.

Multiple Choice

______________ **1.** Water-soluble vitamins include vitamin ___ and a group of vitamins known as the B vitamins.
- A. A
- B. C
- C. D
- D. K

______________ **2.** Antidiuretic hormone (ADH) is a hormone secreted by the body that signals the kidneys to ___.
- A. eliminate water
- B. conserve water
- C. tighten blood vessels
- D. relax blood vessels

______________ **3.** ___ is obtained from preformed retinoid or provitamin carotenoid.
- A. Zinc
- B. Iodine
- C. Vitamin A
- D. Vitamin D

______________ **4.** ___, sodium, and chloride are the three electrolytes that are crucial to balancing water in and around cells.
- A. Potassium
- B. Pantothenic acid
- C. Biotin
- D. Manganese

______________ **5.** A(n) ___ is an unstable compound produced as a by-product of metabolism.
- A. antioxidant
- B. hemoglobin
- C. free radical
- D. enzyme

______________ **6.** ___ is a condition in which the body does not have an adequate amount of water to function properly.
- A. Anemia
- B. Scurvy
- C. Hyponatremia
- D. Dehydration

______________ **7.** ___ is considered the best source of hydration for the body.
- A. Water in its pure form
- B. Tea
- C. Pasteurized milk
- D. Juice

____________________ **8.** Research shows that consuming foods containing nonheme iron with foods containing ___ helps increase nonheme iron absorption.

A. vitamin A
B. vitamin C
C. iodine
D. chromium

____________________ **9.** Although naturally found in most foods, the majority of ___ in the U.S. diet comes from the salt added to processed foods.

A. sodium
B. zinc
C. copper
D. magnesium

____________________ **10.** A trace mineral is a mineral with a recommended dietary allowance of less than ___ mg per day.

A. 100
B. 200
C. 300
D. 400

____________________ **11.** Key sources of zinc include ___, calf liver, red meat, and poultry.

A. fruits
B. vegetables
C. oils
D. oysters

____________________ **12.** ___ can be synthesized by the body after exposure to sunlight.

A. Vitamin C
B. Vitamin D
C. Calcium
D. Fluoride

____________________ **13.** An antioxidant is a substance that may protect cells from damage caused by ___.

A. free radicals
B. phytonutrients
C. electrolytes
D. coenzymes

____________________ **14.** A(n) ___ is a substance that causes an increased output of urine.

A. electrolyte
B. intrinsic factor
C. antidiuretic hormone
D. diuretic

______________ **15.** The feeling of thirst is caused by the functioning of a gland at the base of the brain known as the ___.
A. salivary gland
B. thyroid gland
C. hypothalamus
D. gallbladder

______________ **16.** Vitamin ___ is a fat-soluble vitamin that plays an essential role in blood clotting and bone formation.
A. B_6
B. C
C. D
D. K

______________ **17.** Water and toothpaste with added ___ have been shown to considerably reduce the incidence of tooth decay.
A. fluoride
B. chromium
C. iodine
D. zinc

______________ **18.** ___ is the most abundant mineral in the body.
A. Calcium
B. Phosphorus
C. Potassium
D. Sodium

______________ **19.** The best dietary sources of vitamin C are ___.
A. dairy products
B. eggs
C. vegetables and fruits
D. meats and poultry

______________ **20.** Beriberi is a disease caused by a ___ deficiency, with symptoms that include muscle weakness, fatigue, nerve damage, edema, and heart damage.
A. riboflavin
B. thiamin
C. vitamin A
D. vitamin K

______________ **21.** Large amounts of ___ and most minerals can be lost in hot cooking liquid.
A. vitamins A and D
B. vitamins E and K
C. water-soluble vitamins
D. fat-soluble vitamins

_______________ **22.** ___ is regarded as the most important nutrient.
A. Water
B. Calcium
C. Vitamin C
D. Iron

_______________ **23.** ___ is a protein produced in the stomach that promotes vitamin B_{12} absorption.
A. An electrolyte
B. A coenzyme
C. Pantothenic acid
D. Intrinsic factor

_______________ **24.** For adult males, the adequate intake for water is ___ L per day.
A. 1.7
B. 2.7
C. 3.7
D. 4.7

_______________ **25.** ___ is the second most abundant mineral in the body.
A. Calcium
B. Phosphorus
C. Magnesium
D. Selenium

Completion

_______________ **1.** ___ is the degree to which a substance can harm living organisms.

_______________ **2.** A(n) ___ is a mineral compound that forms positively or negatively charged ions when dissolved in body fluids.

_______________ **3.** ___ is the movement of water across cell membranes from a dilute solution to a more concentrated solution.

_______________ **4.** ___ is a blood condition in which red blood cells do not supply an adequate amount of oxygen to the body.

_______________ **5.** A(n) ___ is the lack of a substance necessary to maintain health.

_______________ **6.** Salt is comprised of approximately 40% sodium and 60% ___.

_______________ **7.** ___ iron is a type of dietary iron found only in animal-based foods and is easily absorbed by the body.

_______________ **8.** A(n) ___ mineral is a mineral with a recommended dietary allowance that exceeds 100 mg per day.

_______________ **9.** A(n) ___ is an organic nutrient that is required in small amounts to help regulate body processes.

_______________ **10.** A(n) ___ is a substance that has the ability to dissolve other substances.

Critical Thinking

Use complete sentences to respond to the following.

1. Explain why water is regarded as the most important of all nutrients.

2. Choose five vitamins and/or minerals and describe diseases and symptoms associated with their deficiencies.

3. Explain how vitamins and minerals react to air, light, heat, moisture, and pH during the preparation of foods.

Water, Vitamins & Minerals

Chapter 6

ACTIVITIES

Name ______________________ Date ______________

Activity 6-1: Water

A person can live for weeks without food, but when deprived of water, a person can survive for only a few days. Water is indispensable to life because of the essential role it plays in a variety of body functions.

Indicate which of the following is an essential role of water by responding "yes" or "no."

______ **1.** Promotes growth and maintenance

______ **2.** Regulates body temperature

______ **3.** Provides energy

______ **4.** Sustains health of tissues, organs, and joints

______ **5.** Produces antibodies

______ **6.** Produces enzymes

______ **7.** Maintains fluid balance

______ **8.** Produces hormones

______ **9.** Promotes blood clotting

______ **10.** Aids digestion and absorption

Match the letter that represents the missing word in each of the following statements.

______ **11.** Water in ___ helps begin digestion in the mouth.

______ **12.** Water promotes ___ of food through the esophagus and digestive tract.

______ **13.** Water breaks down ___ and readies them for transport through the body in the stomach.

______ **14.** Water helps nutrient ___ in the small intestine.

______ **15.** Water facilitates ___ in the large intestine.

______ **16.** Water acts as a(n) ___ to break down and transport food through the digestive tract.

______ **17.** Approximately 66% of the water in the body is ___ and can be found in the cells of muscles, organs, and skin.

______ **18.** The ___ of water protects the body from extreme temperature changes.

______ **19.** Water provides ___ for joints to facilitate smooth, friction-free movement.

______ **20.** Water follows the movement of electrolytes through cell membranes by a process called ___.

A. nutrients
B. solvent
C. transport
D. intracellular
E. excretion
F. osmosis
G. saliva
H. absorption
I. lubrication
J. heat capacity

Activity 6-2: Water Dietary Analysis

Water toxicity is uncommon. When it does occur, electrolyte balance is disturbed, which can lead to muscle cramps, nausea, dizziness, seizures, and coma. In contrast, water deficiency can lead to dehydration, which reduces the amount of oxygen the blood can transport through the body. Symptoms of dehydration can include an elevated heart rate, increased body temperature, muscle spasms, impaired mental functioning, kidney failure, coma, and even death.

Access your online dietary analysis profile that was created in Activity 2-7 from one of the following websites:

FatSecret™: www.fatsecret.com
SelfNutritionData: http://nutritiondata.self.com/mynd/mytracking
SparkPeople®: www.sparkpeople.com
SuperTracker: www.supertracker.usda.gov

1. Submit a copy of your seven-day dietary analysis.
2. Submit a written report that includes the following criteria:
   a. Describe how the daily water intake compares with USDA dietary recommendations for your age, gender, and physical activity level.
   b. Aside from beverages, determine which foods contributed the most to daily water intake.
   c. Identify three changes that can be made to improve water consumption.

Dietary Analysis Rubric

Content	
Copy of dietary analysis	10
Written report	5
Quality content	10
Summary	5
Format	
Typed, double spaced	5
Correct spelling	5
Correct grammar	5
Correct punctuation	5
Total	50

Activity 6-3: Vitamin Functions

Vitamins are organic nutrients required by the body in small amounts to help regulate body processes. Vitamins do not supply energy but are necessary for optimal health. Water-soluble vitamins (vitamin C and the B vitamins) dissolve in water. Fat-soluble vitamins (vitamins A, D, E, and K) dissolve in fat.

Match each vitamin or group of vitamins to its main functions.

__________ **1.** Promotes healthy vision, proper growth and development, and a healthy immune system

__________ **2.** Aids in energy production, growth, and red cell production; activates vitamin B_6

__________ **3.** Aids immune functions; helps produce collagen; protects against cardiovascular disease, certain cancers, and eye disease

__________ **4.** Plays essential role in blood clotting; helps form bone structure

__________ **5.** Metabolizes carbohydrates; regulates nerve impulses; aids in heart and muscle functions

__________ **6.** Helps produce energy; synthesizes fatty acids; assists in nervous system function; promotes healthy skin, hair, and nails

__________ **7.** Helps produce energy and form hemoglobin; aids in nervous system and immune system functions

__________ **8.** Helps kidney functioning; necessary for absorption of vitamin A, calcium, phosphorus, magnesium, iron, and zinc; promotes healthy teeth and bones

__________ **9.** Produces hormones; metabolizes carbohydrates and lipids; helps manufacture red blood cells; helps synthesize cholesterol

__________ **10.** Supports red blood cell production; assists in forming DNA; helps support nervous system; promotes blood circulation

__________ **11.** Helps support formation of red blood cells; promotes proper nerve development; assists in energy production

__________ **12.** Promotes cellular health; widens blood vessels; fights invading bacteria and viruses

__________ **13.** Necessary for skin, nerve, and digestive system functions; aids in energy production

__________ **14.** Required every 2–3 days; stored in liver and fatty tissues

__________ **15.** Required daily; filtered in kidneys; passed in urine

A. Vitamin C
B. Thiamin
C. Riboflavin
D. Niacin
E. Pantothenic acid
F. Vitamin B_6
G. Biotin
H. Folate
I. Vitamin B_{12}
J. Vitamin A
K. Vitamin D
L. Vitamin E
M. Vitamin K
N. Water-soluble vitamins
O. Fat-soluble vitamins

Activity 6-4: Vitamin Dietary Analysis

Vitamins are a powerful group of substances and are necessary for health. Some vitamins are excreted daily, making toxicities rare. However, deficiencies of these vitamins may be more likely if they are not replenished often. In contrast, vitamins that are stored by the body can lead to toxicities when excessive amounts are consumed.

Access your online dietary analysis profile that was created in Activity 2-7 from one of the following websites:

FatSecret™: www.fatsecret.com
SelfNutritionData: http://nutritiondata.self.com/mynd/mytracking
SparkPeople®: www.sparkpeople.com
SuperTracker: www.supertracker.usda.gov

1. Submit a copy of your seven-day dietary analysis.
2. Submit a written report that includes the following criteria:
   a. Describe how the daily vitamin intake for vitamin C, thiamin, riboflavin, niacin, pantothenic acid, vitamin B_6, biotin, folate, vitamin B_{12}, vitamin A, vitamin D, vitamin E, and vitamin K compares with USDA dietary recommendations for your age, gender, and physical activity level.
   b. Determine which foods contributed the greatest amounts of the identified vitamins.
   c. Identify three changes that can be made to improve vitamin consumption.

Dietary Analysis Rubric

Content	
Copy of dietary analysis	10
Written report	5
Quality content	10
Summary	5
Format	
Typed, double spaced	5
Correct spelling	5
Correct grammar	5
Correct punctuation	5
Total	50

Activity 6-5: Mineral Functions

Minerals are inorganic nutrients required by the body in small amounts to help regulate body processes. Minerals occur naturally in soil and in the water absorbed by plants, making a variety of minerals plentiful in plant-based foods.

Match each mineral to its main functions.

__________ **1.** Required to make oxygen-carrying proteins hemoglobin and myoglobin

__________ **2.** Protects teeth against tooth decay; helps with formation of healthy teeth and bones

__________ **3.** Helps maintain fluid and acid-base balance

__________ **4.** Required for healthy teeth and bones; necessary for muscle contraction; helps carry messages from the brain to all body parts; assists in blood clotting

__________ **5.** Helps regulate normal heart functioning and muscle contraction; necessary for energy production and protein synthesis; helps maintain fluid and acid-base balance

__________ **6.** Helps form red blood cells; involved in enzymatic reactions that keep tissues, nerves, and bones healthy

__________ **7.** Keeps bones strong; helps facilitate many chemical reactions that are necessary for healthy immune system, proper nerve and muscle functioning, and maintaining normal blood glucose and normal blood pressure; helps metabolize proteins and lipids

__________ **8.** Required to make hormones in the thyroid gland

__________ **9.** Works with calcium to build strong teeth and bones; helps synthesize proteins; assists in energy production and storage; filters waste from kidneys; helps form DNA

__________ **10.** Works as coenzyme to facilitate protein and carbohydrate metabolism and bone formation; helps maintain nerve functions; promotes synthesis of fatty acids and cholesterol

__________ **11.** Helps regulate blood glucose levels

__________ **12.** Helps maintain thyroid hormone functions; assists in keeping immune system healthy

__________ **13.** Helps protect immune system; plays a role in regulating genetic material; promotes proper functioning of senses; assists in wound healing, protein synthesis, energy metabolism, and insulin storage

__________ **14.** Helps maintain acid-base balance, proper blood volume, and blood pressure

A. Calcium
B. Phosphorus
C. Magnesium
D. Potassium
E. Sodium
F. Chloride
G. Iron
H. Zinc
I. Iodine
J. Selenium
K. Copper
L. Manganese
M. Fluoride
N. Chromium

Activity 6-6: Mineral Dietary Analysis

Minerals are essential for optimal health and well-being. Toxic reactions to minerals can happen quickly, or they can build over time. Consuming excessive amounts of one mineral can also cause a deficiency of another mineral. Adverse symptoms that range in severity may occur when the body is deficient in a mineral.

Access your online dietary analysis profile that was created in Activity 2-7 from one of the following websites:

FatSecret™: www.fatsecret.com
SelfNutritionData: http://nutritiondata.self.com/mynd/mytracking
SparkPeople®: www.sparkpeople.com
SuperTracker: www.supertracker.usda.gov

1. Submit a copy of your seven-day dietary analysis.
2. Submit a written report that includes the following criteria:
   a. Describe how the daily mineral intake for calcium, phosphorus, magnesium, potassium, sodium, chloride, iron, zinc, iodine, selenium, copper, manganese, fluoride, and chromium compares with USDA dietary recommendations for your age, gender, and physical activity level.
   b. Determine which foods contributed the greatest amounts of the identified minerals.
   c. Identify three changes that can be made to improve mineral consumption.

Dietary Analysis Rubric

Content	
Copy of dietary analysis	10
Written report	5
Quality content	10
Summary	5
Format	
Typed, double spaced	5
Correct spelling	5
Correct grammar	5
Correct punctuation	5
Total	50

Activity 6-7: Water Content in Foods Lab

Objective: To measure the amount of water contained in carrots, celery, cucumbers, and dark, leafy greens.

Materials and Equipment

- Gram scale
- 3 carrots
- 3 celery stalks
- 1 cucumber
- 3 cups spinach, kale, or other dark, leafy green
- Cutting board
- Chef's knife
- 8 bowls
- Juicer
- Rubber spatula
- Cheesecloth
- Fine mesh strainer
- Calculator

Procedure

1. Weigh the carrots and record their weight.
2. Cut the carrots into pieces that will fit in the juicer.
3. Position a bowl by the juicer so it will collect the liquid extracted from the carrots.
4. Juice the carrots and collect the liquid. Turn off the juicer.
5. Scrape all the residue that remains inside the juicer onto two layers of cheesecloth.
6. Squeeze the cheesecloth over the bowl that contains the carrot juice to extract as much liquid as possible. Discard the cheesecloth.
7. Set an empty bowl on the scale and set (tare) the scale to 0. Remove the bowl.
8. Place a mesh strainer over the empty bowl and pour the carrot juice into the strainer.
9. Press any solids that remain in the strainer with a rubber spatula to extract the liquid.
10. Weigh the bowl of juice.
11. Divide the weight of the juice by the original weight of the carrots to determine the percentage of water contained in the carrots. Record the results.
12. Repeat steps 1–11 for the celery stalks.
13. Repeat steps 1–11 for the cucumber.
14. Repeat steps 1–11 for the spinach, kale, or other dark, leafy green.

Complete the following chart.

	Carrots	Celery	Cucumber	Dark, Leafy Greens
Food weight				
Juice weight				
Water percentage (juice weight ÷ food weight)				

Activity 6-8: Vitamin and Mineral Supplements Research

Individuals ranging in age from young children to the elderly commonly take vitamin and/or mineral supplements. While vitamin and mineral supplements can be beneficial, they can also involve health risks.

Access an article from a medical-based website, a nutrition journal, and a company that sells vitamin and mineral supplements. Prepare a three- to four-page report and an oral presentation that address the following criteria.

1. Explain how vitamin and mineral supplements are regulated.
2. Identify the vitamins and minerals commonly found in supplements.
3. Describe deficiencies that can be prevented or treated by taking vitamin and mineral supplements.
4. Describe toxicities that can occur by taking vitamin and mineral supplements.
5. Explain the role of vitamin and mineral supplements when combined with a well-balanced diet.
6. Explain how this research will affect your selection of vitamin and mineral supplements in the future.

Nutrition Project Rubric

Content	
Written report	5
Cover page	5
Introduction	5
Quality content	20
Summary	5
Format	
Typed, double spaced	5
Correct spelling	5
Correct grammar	5
Correct punctuation	5
Bibliography	5
Oral Presentation	
Introduction	5
Quality content	10
Summary	5
Delivery	5
Visual aid(s)	10
Total	100

Nutritious Menu Planning

Chapter 7

REVIEW

Name ______________________________ **Date** ____________________

True-False

T	F	**1.** The "farm to table" movement focuses on increasing the transparency between where food is grown and where it is consumed.
T	F	**2.** When purchasing ingredients, emphasis should be placed on selecting high-quality whole foods that have undergone the least amount of processing possible.
T	F	**3.** Menu items should appeal to only two senses: taste and smell.
T	F	**4.** When a recipe is modified, cooking times may change.
T	F	**5.** Beverages are usually listed last on the menu or presented on a separate menu.
T	F	**6.** Foods grown using a sustainable philosophy are produced in a manner that does not harm the environment.
T	F	**7.** Research indicates that the terms "low-fat" and "low-carb" are the most effective menu descriptors.
T	F	**8.** Menu promotions are commonly used to generate business during slow times.
T	F	**9.** Inaccurate menu labeling can jeopardize the health of guests.
T	F	**10.** Using ready-to-use soup stock instead of house-made soup stock is an effective way to reduce sodium levels.
T	F	**11.** A menu description should communicate both the main ingredients of a dish and its method of preparation.
T	F	**12.** The truth-in-menu guidelines govern only written menu descriptions.
T	F	**13.** If nutrient claims are made, the nutrition information must appear on the menu.
T	F	**14.** Research suggests that flavor is the most important factor when selecting foods.
T	F	**15.** Advantages of using cycle menus include ease in scheduling staff, ordering products, and controlling inventory.

Multiple Choice

_______________ **1.** A ___ menu is a menu that changes frequently to coincide with changes in product availability.
A. cycle
B. fixed
C. market
D. prix fixe

_______________ **2.** Truth-in-menu guidelines were established by the federal government to ___.
A. ensure the accuracy of menu statements
B. ensure the quality of menu items
C. standardize the format of menus
D. promote healthy menu items

_______________ **3.** The FDA describes a C-ranked health claim as follows: ___.
A. Scientific evidence supports the claim
B. "...Although there is scientific evidence supporting the claim, the evidence is not conclusive"
C. "Very limited and preliminary scientific research suggests..."
D. "Some scientific evidence suggests..."

_______________ **4.** The first step of the culinary nutrition recipe-modification process is to ___.
A. evaluate the original recipe for sensory and nutritional qualities
B. establish goals for the recipe modification
C. select appropriate modifications or substitutions
D. test the modified recipe

_______________ **5.** On a single-page menu, the eye naturally falls to the ___ of the page, which is an ideal location for marketing nutritious menu items.
A. left
B. right
C. center
D. bottom

_______________ **6.** A(n) ___ menu is a menu that identifies specific items that will be served for each course at a set price.
A. à la carte
B. prix fixe
C. California
D. table d'hôte

_______________ **7.** ___ venues provide seated guests with individualized service, refined presentations, and a sophisticated environment.
A. Fast casual
B. Casual dining
C. Fine dining
D. Institutional

_______________ **8.** A ___ menu is a menu that offers all food and beverage items for breakfast, lunch, and dinner throughout the entire day.
- A. cycle
- B. California
- C. prix fixe
- D. table d'hôte

_______________ **9.** The Patient Protection and Affordable Care Act of 2010 requires foodservice operations with ___ or more locations to list the calories for standard menu items on restaurant menus and menu boards.
- A. 5
- B. 10
- C. 15
- D. 20

_______________ **10.** ___ venues, such as educational, health care, and correctional facilities, generally serve large numbers of people.
- A. Institutional
- B. Special event
- C. Fine dining
- D. Quick service

Completion

_______________ **1.** ___ is the amount of a particular product that should be kept in inventory to ensure that an adequate supply is on hand for normal production.

_______________ **2.** A(n) ___ is the assortment of items that may be ordered from a given menu.

_______________ **3.** An ingredient ___ is an ingredient that replaces an item of similar characteristics.

_______________ **4.** A(n) ___ is a written form listing the specific characteristics of a product that is to be purchased from a supplier.

_______________ **5.** A(n) ___ is a statement declaring a relationship between certain foods or nutrients and disease.

_______________ **6.** A(n) ___ menu is a menu that offers separately priced food and beverages.

_______________ **7.** A(n) ___ is a document that markets a foodservice operation to guests.

_______________ **8.** Nutrient and health claims that appear on menus and food labels must follow requirements and definitions established by the ___.

_______________ **9.** A(n) ___ is a food that is larger than a single bite and is typically served as the first course of a meal.

_______________ **10.** An ingredient ___ is an ingredient that replaces an item of different characteristics.

Critical Thinking

Use complete sentences to respond to the following.

1. Describe current menu trends and how they may impact foodservice operations.

2. Explain how to use flavor development, cooking techniques, sensory qualities, and plate composition to dispel the misconception that "healthy" food lacks flavor.

3. Explain how to effectively use menu descriptions, menu item placement, perceived value, and menu promotions to market nutritious menu items.

Nutritious Menu Planning

Chapter 7

ACTIVITIES

Name ______________________________ **Date** ________________

Activity 7-1: Menu Modification

Foodservice operations often provide nutrition information for menu items. The ability to identify menu items that are high in calories, fat, sugar, and/or sodium and modify those items to create more nutrient-dense selections is an essential skill that helps meet the growing demand for healthy menu options.

Using a menu that lists nutrition information, respond to the following.

1. Select a menu item and list the nutrition information provided for that item.

2. List the goals for modifying the menu item.

3. Explain how to modify the menu item to achieve the listed goals.

4. Describe how the modified menu item might be promoted.

Activity 7-2: Menu Descriptions

Menu descriptions should be clear and appealing while communicating the main ingredients of a dish and the method of preparation. Effective menu descriptions promote clarity, add appeal, and increase the likelihood that the item will be ordered.

Ineffective Menu Description: Veggie Burger

Effective Menu Description: Grilled portobello-lentil burger with chipotle-lime yogurt on a toasted whole grain bun

Rewrite the following ineffective menu descriptions to make them more appealing and complete.

1. Oatmeal

2. Chef's Salad

3. Steak Fajita

4. Shrimp Kebab

5. Gazpacho

Menu Descriptions Rubric					
Description Criteria	**Oatmeal**	**Chef's Salad**	**Steak Fajita**	**Shrimp Kebab**	**Gazpacho**
Accurate representation	1	1	1	1	1
Promotes clarity	1	1	1	1	1
Appealing portrayal	1	1	1	1	1
States unique characteristics	1	1	1	1	1
Conveys preparation techniques	1	1	1	1	1

Scoring: A point is earned for each menu item description that meets the criteria indicated in the rubric for a possible total score of 25 points.

Activity 7-3: Bologness Sauce Recipe Modification

The purpose of the culinary nutrition recipe-modification process is to increase the nutrient density of a menu item as well as to accommodate guests with dietary considerations such as food allergies. The appearance, texture, and flavor of menu items should also be considered when modifying recipes.

Modify the following Bolognese sauce recipe to reduce its calories, fat, and sodium.

Yield: 8 servings, 1½ cups each

Ingredients

⅓ c	olive oil
1½ lb	ground beef
6 ea	garlic cloves, thinly sliced
1 c	onion, small dice
½ c	celery, chopped
½ c	carrot, chopped
¼ c	Italian parsley, chopped
2 ea	plum tomatoes, 28 oz canned
2 tsp	salt
1 tsp	black pepper, ground

Preparation

1. Heat oil in a pan over medium heat. Add ground beef and cook until beef is no longer pink.
2. Add garlic, onion, celery, and carrot and cook until the vegetables soften.
3. Add parsley, tomatoes, salt, and pepper. Stir to combine.
4. Allow sauce to simmer for 45 minutes.

Culinary Nutrition Recipe-Modification Process

Step 1: Evaluate original recipe for sensory and nutritional qualities.

Step 2: Establish goals for recipe modifications.

Step 3: Identify modifications or substitutions.

Step 4: Determine functions of identified modifications or substitutions.

Step 5: Select appropriate modifications or substitutions.

Step 6: Test modified recipe to evaluate sensory and nutritional qualities.

Activity 7-4: Menu Item Ingredient Substitutions Lab

An ingredient substitution is an ingredient that replaces an item of similar characteristics. Ingredient substitutions are commonly used to reduce the fat, sugar, and/or sodium content of a dish.

Objective: To evaluate the texture and flavor of products used as ingredient substitutions to reduce the fat, sugar, or sodium content.

Materials and Equipment

- Marker
- 6 tasting cups
- 6 tasting spoons
- Can opener
- Food items
  - sour cream
  - low-fat sour cream
  - peaches canned in syrup
  - peaches canned in water
  - chicken broth
  - low-sodium chicken broth

Procedure

1. Label two tasting cups, one for each type of sour cream.
2. Place a tasting portion of the sour cream and the low-fat sour cream into the correctly labeled cup.
3. Taste and evaluate the sour cream and the low-fat sour cream in terms of texture and flavor.
4. Repeat steps 1–3 for the peaches.
5. Repeat steps 1–3 for the chicken broths.

Respond to the following.

1. Compare the texture and flavor of sour cream and low-fat sour cream.

2. Identify a menu item where low-fat sour cream could be used to reduce fats.

3. Compare the texture and flavor of peaches canned in syrup and peaches canned in water.

4. Identify a menu item where peaches canned in water could be used to reduce sugars.

5. Compare the texture and flavor of chicken broth and low-sodium chicken broth.

6. Identify a menu item where low-sodium chicken broth could be used to reduce sodium.

Activity 7-5: Seasoning Blends Research

Chefs use signature seasoning blends to develop flavorful menu items. These blends usually enhance flavor while limiting calories.

Access the USDA National Nutrient Database for Standard Reference website at http://ndb.nal.usda.gov/. Prepare a two- to three-page report and an oral presentation that address the following criteria.

1. Create a signature seasoning blend.
2. Describe the intended flavor profile of the seasoning blend and list the ingredients.
3. Identify key nutrients found in the seasoning blend.
4. Prepare a food item with the seasoning blend and describe the results in terms of flavor.
5. Identify three changes that could be made to the original recipe to improve nutrient density and flavor.

Nutrition Project Rubric

Content	
Written report	5
Cover page	5
Introduction	5
Quality content	20
Summary	5
Format	
Typed, double spaced	5
Correct spelling	5
Correct grammar	5
Correct punctuation	5
Bibliography	5
Oral Presentation	
Introduction	5
Quality content	10
Summary	5
Delivery	5
Visual aid(s)	10
Total	100

Beverages on the Menu

Chapter 8

REVIEW

Name ______________________________ **Date** ____________________

True-False

T F **1.** Because soft drinks are mainly comprised of refined sugars, they lack nutrients and are high in empty calories.

T F **2.** Fruit and vegetable beverages are a healthy source of vitamins and minerals and do not contain sugar.

T F **3.** The cost of purchasing and preparing beverages is generally low in comparison to beverage menu prices.

T F **4.** Buttermilk and kefir contain bacteria found to impair digestive health.

T F **5.** Soy milk supplies approximately 8 g of protein per cup but is high in saturated fat.

T F **6.** Regular coffee has 80–200 mg of caffeine per 8 fl oz.

T F **7.** Black tea is graded according to the size of its leaves.

T F **8.** Rice milk is a hypoallergenic beverage choice.

T F **9.** White teas undergo the most processing of any tea.

T F **10.** The liver, stomach, and kidneys are directly affected by alcohol consumption.

T F **11.** Diet soft drinks are popular zero-calorie beverages sweetened with natural sugars.

T F **12.** Milklike beverages usually contain gluten.

T F **13.** Serving alcohol illegally can result in fines, loss of a liquor license, imprisonment, and/or the closing of a business.

T F **14.** Sparkling bottled water is obtained from spring water that is naturally carbonated.

T F **15.** The width of the glass does not affect the color and head of the beer.

Multiple Choice

_______________ **1.** Milklike beverages are not dairy products and do not contain ___.

A. lactose
B. calcium
C. fat
D. calories

_______________ **2.** In general, ___ of ground coffee is used for every 6–7 fl oz of water.

A. 1 tsp
B. 2 tsp
C. 1 tbsp
D. 2 tbsp

_______________ **3.** Although derived from a fruit, not a nut, ___ milk is often categorized with nut-based milklike beverages.

A. hemp
B. peanut
C. coconut
D. hazelnut

_______________ **4.** ___ water is purified water that has been boiled until it steams and then cooled until it condenses back to liquid form.

A. Artesian
B. Distilled
C. Mineral
D. Spring

_______________ **5.** ___ does not contain calories and is considered the best source of hydration for the body.

A. Pure drinking water
B. Flavored water
C. Tea
D. Fruit juice

_______________ **6.** Green drinks (green smoothies) are high in dietary fiber and typically low in ___.

A. water-soluble vitamins
B. fat-soluble vitamins
C. sugar
D. antioxidants

_______________ **7.** ___ is an intensely flavored coffee made from beans that have been roasted to the very dark stage.

A. Espresso
B. Robusta
C. Arabica
D. Vienesse roast

_______________ **8.** The grade of tea refers to the ___ of its leaves.
A. flavor
B. color
C. quality
D. size

_______________ **9.** The American Academy of Pediatrics recommends that adolescents consume no more than ___ mg of caffeine per day.
A. 50
B. 100
C. 150
D. 200

_______________ **10.** A sommelier is an individual formally trained in ordering, storing, and serving ___.
A. coffees
B. teas
C. wines
D. beers

_______________ **11.** To properly brew tea, the water is heated to just below the boiling point, and the tea is steeped for ___ minutes.
A. 1–2
B. 3–5
C. 6–7
D. 9–11

_______________ **12.** Proof is ___ the percentage of alcohol by volume.
A. a quarter
B. half
C. twice
D. three times

_______________ **13.** Juices made from 100% juice are labeled "___" or "100% juice."
A. juice
B. fruit drink
C. juice cocktail
D. fruit cocktail

_______________ **14.** Beer is ___% water, and the quality of the water greatly affects the finished product.
A. 20
B. 40
C. 70
D. 90

_______________ **15.** Carbonated water is water that has been infused with ___.
- A. carbon dioxide
- B. carbon monoxide
- C. oxygen
- D. chloride

_______________ **16.** ___ is calorie-free, contains a low to moderate amount of caffeine, and provides a variety of antioxidants.
- A. Juice
- B. Tea
- C. Sake
- D. A diet soft drink

_______________ **17.** A longer roast produces a darker-colored coffee bean with a ___ flavor.
- A. sweeter
- B. saltier
- C. more subtle
- D. more robust

_______________ **18.** ___ wines are served in tall, narrow flutes to retain carbonation.
- A. Fortified
- B. Sparkling
- C. Red
- D. White

_______________ **19.** ___ wines have been supplemented with spirits such as brandy and may also contain aromatic herbs and spices.
- A. Still
- B. Sparkling
- C. Vintage
- D. Fortified

_______________ **20.** The amount of caffeine found in soft drinks, such as cola, is limited by the FDA to ___ mg per 12 fl oz.
- A. 53
- B. 71
- C. 96
- D. 117

Completion

_______________ **1.** A(n) ___ beer is produced by a small, independently owned brewer.

_______________ **2.** A(n) ___ is a carbonated beverage combined with ice cream.

_______________ **3.** A(n) ___ is an herbal beverage created by steeping herbs, spices, flowers, dried fruits, or roots in boiling water.

_______________ 4. A(n) ___ is an alcoholic beverage made from distilled alcohol and flavored with fruits, spices, herbs, flowers, nuts, or cream, and bottled with added sugar.

_______________ 5. A(n) ___ is an astringent compound found in tea, the bark of some trees, and the skins, seeds, and stems of grapes.

_______________ 6. ___ is a Japanese wine made from rice and is not aged.

_______________ 7. A(n) ___ is an alcoholic beverage commonly made from fermented fruit juice that has been aged for a period of time.

_______________ 8. ___ water is bottled still water obtained by tapping an underground water source, which causes the water to rise into a well.

_______________ 9. ___ wine is often called Champagne, but only Champagne that originates from the Champagne region in France can be labeled as such.

_______________ 10. ___ is a beer, such as a porter or a stout, that is fermented rapidly.

Critical Thinking

Use complete sentences to respond to the following.

1. Compare still, carbonated, flavored, and enhanced water.

2. Summarize the nutritional qualities of four milklike beverages.

3. Explain why soft drink consumption can be a nutritional concern and include one solution to combat this concern.

4. Describe the types of beverages commonly served during breakfast, lunch, dinner, and dessert.

5. Identify ways to promote beverages on the menu.

Beverages on the Menu

Chapter 8

ACTIVITIES

Name ______________________________ **Date** ____________________

Activity 8-1: Nutrients in Beverages

Many beverages supply essential nutrients such as carbohydrates, protein, vitamins, and minerals. However, beverages can also be high in calories, fat, sugar, caffeine, and artificial ingredients, making portion control essential. For quenching thirst and hydrating the cells of the body, water is the best beverage.

Identify the item from each of the following pairs that is lowest in calories.

__________________ **1.** nonfat milk / soy milk

__________________ **2.** cappuccino, low-fat / coffee, regular

__________________ **3.** orange juice / carrot juice

Identify the item from each of the following pairs that is highest in calories.

__________________ **4.** flavored water / carbonated water

__________________ **5.** white wine / red wine

__________________ **6.** regular beer / light beer

Identify the item from each of the following pairs that is highest in fat.

__________________ **7.** 2% milk / chocolate milk

__________________ **8.** cappuccino, low-fat / coffee, regular

__________________ **9.** orange juice / tomato juice

Identify the item from each of the following pairs that is highest in carbohydrates.

__________________ **10.** flavored water / carbonated water

__________________ **11.** brewed tea / iced tea, sweetened

__________________ **12.** tomato juice / carrot juice

Identify the item from each of the following pairs that is lowest in protein.

__________________ **13.** nonfat milk / chocolate milk

__________________ **14.** cappuccino, low-fat / coffee, regular

__________________ **15.** tomato juice / carrot juice

Activity 8-2: Beverage Menu Descriptions

Menu descriptions should be clear and appealing while communicating the main ingredients of a dish and the method of preparation. Effective menu descriptions promote clarity, add appeal, and increase the likelihood that the item will be ordered.

Rewrite the following ineffective menu descriptions to make them more appealing and complete.

1. Fruit Smoothie

2. Lactose-Free Cappuccino

3. Herbal Beverage

4. Craft Beer

5. Bloody Mary

Menu Descriptions Rubric					
Description Criteria	**Fruit Smoothie**	**Lactose-Free Cappuccino**	**Herbal Beverage**	**Craft Beer**	**Bloody Mary**
Accurate representation	1	1	1	1	1
Promotes clarity	1	1	1	1	1
Appealing portrayal	1	1	1	1	1
States unique characteristics	1	1	1	1	1
Conveys preparation techniques	1	1	1	1	1

Scoring: A point is earned for each menu item description that meets the criteria indicated in the rubric for a possible total score of 25 points.

Activity 8-3: Piña Colada Recipe Modification

The purpose of the culinary nutrition recipe-modification process is to increase the nutrient density of a menu item as well as to accommodate guests with dietary considerations such as food allergies. The appearance, texture, and flavor of menu items should also be considered when modifying recipes.

Modify the following piña colada recipe to reduce its alcohol, calories, fat, and sugar and increase its dietary fiber, vitamins, and minerals.

Yield: 1 serving, 12 fl oz

Ingredients

1 oz	light rum
1 oz	amber rum
1 oz	dark rum
3 oz	pineapple juice
2 oz	sweet and sour mix
2 oz	cream of coconut
1 oz	heavy cream
½ c	ice cubes

Preparation

1. Place all ingredients in a blender.
2. Blend ingredients to a slush-like consistency.
3. Pour into a glass and serve.

Culinary Nutrition Recipe-Modification Process

Step 1: Evaluate original recipe for sensory and nutritional qualities.

Step 2: Establish goals for recipe modifications.

Step 3: Identify modifications or substitutions.

Step 4: Determine functions of identified modifications or substitutions.

Step 5: Select appropriate modification or substitutions.

Step 6: Test modified recipe to evaluate sensory and nutritional qualities.

Activity 8-4: Aftertaste of Sweeteners in Beverages Lab

Sugars and sugar substitutes are frequently added to beverages. In addition to providing sweetness, some sweeteners leave a detectable aftertaste.

Objective: To determine if a sugar or sugar substitute leaves an aftertaste in a beverage.

Materials and Equipment

- Measuring cup
- 3 qt water
- Saucepan
- 10 teabags
- 12 cups
- Teaspoon measure
- Sugars
  - agave nectar
  - brown sugar
  - corn syrup
  - granulated sugar
  - honey
  - maple syrup
  - molasses
  - turbinado sugar
- Sugar substitutes
  - aspartame
  - saccharin
  - sucralose
  - stevia
- 12 spoons or stirrers

Allergen Alert

The materials used in this activity may contain one or more food allergens.

Procedure

1. Fill a saucepan with 3 qt of water and heat to 180°F.
2. Remove the saucepan from the heat and add teabags. Steep for 3–5 minutes.
3. Remove teabags and place the saucepan over low heat to keep warm.
4. Pour 8 oz of hot tea into a cup. Add 1 tsp of agave nectar and stir until dissolved.
5. Taste the hot tea and evaluate whether the sweetener leaves an aftertaste.
6. Repeat steps 4–5 for all sugars and sugar substitutes.

Respond to the following.

1. Identify the sweeteners that leave an aftertaste.

2. Describe the aftertaste left by these sweeteners.

Activity 8-5: Energy Drinks Research

Energy drinks contain ingredients that act as stimulants to increase energy. While energy drinks are generally considered safe, moderate consumption is recommended.

Access an article from a medical-based website, a nutrition journal, and a company that sells energy drinks. Prepare a three- to four-page report and an oral presentation that address the following criteria.

1. Describe five ingredients commonly found in energy drinks.
2. Describe three health issues related to the consumption of energy drinks.
3. Compare the caffeine content in three energy drinks.
4. Determine whether there are any benefits to energy drinks.
5. Describe a scenario where energy drinks could be dangerous.
6. Explain how this research will affect your selection of energy drinks in the future.

Nutrition Project Rubric

Content	
Written report	5
Cover page	5
Introduction	5
Quality content	20
Summary	5
Format	
Typed, double spaced	5
Correct spelling	5
Correct grammar	5
Correct punctuation	5
Bibliography	5
Oral Presentation	
Introduction	5
Quality content	10
Summary	5
Delivery	5
Visual aid(s)	10
Total	100

Egg, Soy & Dairy Products on the Menu

Chapter 9 REVIEW

Name ______________________________ **Date** ______________________

True-False

T F **1.** An egg consists of four main parts: the shell, shell membrane, yolk, and white.

T F **2.** Silken tofu is solid and dense.

T F **3.** Eggs are a complete protein.

T F **4.** Cream tolerates heat better than milk due to its high fat content.

T F **5.** Soybeans are higher in total fat than other plant-based foods.

T F **6.** Cheese contains saturated fat but no cholesterol.

T F **7.** Edamame is only served raw.

T F **8.** Dairy products are generally rich in calcium.

T F **9.** Eggs can be cooked using both dry-heat and moist-heat cooking techniques.

T F **10.** Eggs with brown shells differ in terms of nutrition and flavor from eggs with white shells.

T F **11.** Soy products such as tofu naturally absorb the flavor of the ingredients they are paired with.

T F **12.** Milk and cream classifications are generally determined by butterfat content.

T F **13.** Butter is full of unsaturated fat.

T F **14.** Egg whites are commonly used to emulsify salad dressings.

T F **15.** Soy products can offer health benefits over animal-based proteins in terms of dietary fiber and cholesterol content.

Multiple Choice

_______________ **1.** The majority of the protein in an egg is found in the ___, which contains 4 g of protein.
- A. shell membrane
- B. egg yolk
- C. egg white
- D. chalazae

_______________ **2.** ___ is a flat, dense cake made from fermented soybeans.
- A. Yuba
- B. Edamame
- C. Natto
- D. Tempeh

_______________ **3.** A(n) ___ is a live microorganism that has been found to be beneficial to the digestive tract.
- A. antibiotic
- B. probiotic
- C. phytonutrient
- D. isoflavone

_______________ **4.** ___ eggs are the standard size used in foodservice operations.
- A. Small
- B. Medium
- C. Large
- D. Jumbo

_______________ **5.** Casein is the main group of ___ found in milk.
- A. proteins
- B. fats
- C. sugars
- D. starches

_______________ **6.** Items made from milk that have little to no calcium, such as ___, are often grouped with dairy products because they are made from milk.
- A. yogurt
- B. kefir
- C. cheese
- D. butter

_______________ **7.** ___ is the curd of ground, cooked soybeans that has been pressed into a cake.
- A. Okara
- B. Tofu
- C. Edamame
- D. Textured soy protein

__________ **8.** ___ does not hold its shape well and is better puréed or blended into soups, sauces, dips, salad dressings, and puddings.

A. Edamame
B. Tempeh
C. Silken tofu
D. Firm tofu

__________ **9.** Overcooking eggs can cause the yolk of a(n) ___ to form a green outer ring.

A. egg cooked in the shell
B. sunny-side up egg
C. shirred egg
D. egg substitute

__________ **10.** Cooking eggs to an internal temperature of ___°F helps destroy potentially harmful bacteria like salmonella enteritidis.

A. 100
B. 120
C. 140
D. 160

__________ **11.** ___ is a semisolid dairy product that has been fermented by adding strains of active bacteria and yeast to milk.

A. Kefir
B. Sour cream
C. Cheese
D. Buttermilk

__________ **12.** ___ is a type of Asian flavoring made from fermented soybeans, starch such as barley or rice, salt, and water.

A. Soy sauce
B. Natto
C. Kecap
D. Miso

__________ **13.** ___ butter is made by mixing cold, softened butter with flavoring ingredients such as herbs.

A. Compound
B. Clarified
C. Complex
D. Brown

__________ **14.** The ___ is the thin, skinlike material located directly under the eggshell.

A. egg yolk
B. egg white
C. shell membrane
D. albumen

__________ **15.** Field soybeans are used to produce ___.
A. condiments
B. soybean oil
C. tofu
D. meat substitutes

__________ **16.** A(n) ___ is an organic compound found in legumes such as soybeans and is considered a phytonutrient.
A. bacterial culture
B. isoflavone
C. polyphenol
D. probiotic

__________ **17.** Most dairy products contain lactose, the primary ___ in milk.
A. protein
B. fat
C. sugar
D. acid

__________ **18.** ___ contains a minimum of 80% butterfat.
A. Cheese
B. Butter
C. Acidophilus milk
D. Cream

__________ **19.** The yolk of one egg contains approximately ___ g of total fat.
A. 2
B. 3
C. 5
D. 8

__________ **20.** The ___ is the watery part of milk.
A. butterfat
B. casein
C. curd
D. whey

__________ **21.** Soy sauce is an Asian sauce made from fermented soybeans, ___, salt, and water.
A. garlic
B. sugar
C. wheat
D. barley

__________ **22.** Greek yogurt has more ___ than regular yogurt.
A. protein
B. carbohydrates
C. sodium
D. calcium

____________________ **23.** A ___ cheese, also known as a rind-ripened cheese, is a cheese that has been sprayed with a harmless live mold to produce a thin skin or rind.

A. fresh
B. soft
C. hard
D. grating

____________________ **24.** Research suggests that individuals following a low-fat and/or low-cholesterol diet can consume ___ eggs per day without experiencing a rise in blood cholesterol levels.

A. 1–2
B. 2–3
C. 3–4
D. 4–5

____________________ **25.** ___ is a granular, meatless protein product made from soy flour.

A. Silken tofu
B. A meat analog
C. Textured soy protein
D. Tempeh

____________________ **26.** ___ are used to bind, emulsify, thicken, coat, clarify, and leaven foods.

A. Eggs
B. Soy products
C. Liquid dairy products
D. Semisolid dairy products

____________________ **27.** Evaporated milk is produced by removing 60% of the ___ from milk.

A. butterfat
B. casein
C. lactose
D. water

____________________ **28.** A(n) ___ is an antioxidant that can cause curding.

A. polyphenol
B. probiotic
C. phytonutrient
D. isoflavone

____________________ **29.** Yuba and okara are ___ by-products used in culinary preparations.

A. edamame
B. tofu
C. soy milk
D. meat analog

____________________ **30.** ___ are one of the few foods considered an excellent source of choline.
A. Eggs
B. Soy products
C. Yogurts
D. Cheeses

Completion

____________________ **1.** ___ are green soybeans.

____________________ **2.** ___ are twisted cordlike strands of egg white that anchor the yolk to the center of the egg white.

____________________ **3.** ___ is the process of emulsifying the fat particles in milk in order to provide a uniform consistency and prevent fat separation.

____________________ **4.** A(n) ___ is a meatless product made from soybean by-products that is formed into the familiar shape of a processed meat product.

____________________ **5.** ___ is a fermented liquid dairy product made by adding bacterial cultures to low-fat or skim milk to create a thickened texture and tangy flavor.

____________________ **6.** ___ soybeans are generally bred for their large seed size, appealing texture and flavor, and short cooking time.

____________________ **7.** ___ is the process of destroying harmful microorganisms by heating and then quick-cooling a food product.

____________________ **8.** ___ is a solid dairy product made from milk curds.

____________________ **9.** The outside of the shell of a freshly laid egg is covered by a protective layer called the ___.

____________________ **10.** The ___ is the thick, casein-rich part of coagulated milk.

____________________ **11.** ___ is a process in which a hot liquid is gradually added to eggs in order to slowly raise the temperature of the eggs, which stabilizes their texture and prevents curdling.

____________________ **12.** ___ is the liquid expressed from soybeans that have been soaked and finely ground.

____________________ **13.** A(n) ___ is an oil-rich seed grown primarily for its oil.

____________________ **14.** The egg white, also known as the ___, is the clear portion of a raw egg.

____________________ **15.** A(n) ___-rind cheese is a semisoft cheese that is allowed to ripen through exposure to air.

Critical Thinking

Use complete sentences to respond to the following.

1. Describe the nutritional qualities of eggs.

2. Explain the role of eggs in beverages, appetizers, soups, sandwiches, salads, entrées, and desserts.

3. Summarize the benefits of soy products in terms of nutrition, cost, and shelf life.

4. Describe how soy products can be incorporated into beverages, appetizers, soups, salads, entrées, and desserts.

5. Choose three types of dairy products and explain how they are effectively cooked and used to prepare menu items.

6. Explain the role of dairy products in beverages, appetizers, soups, sandwiches, salads, entrées, sides, and desserts.

Egg, Soy & Dairy Products on the Menu

Chapter 9

ACTIVITIES

Name ______________________________ **Date** ____________________

Activity 9-1: Egg, Soy, and Dairy Products

Egg, soy, and dairy products are often used to add texture and flavor, but they also provide protein, essential fatty acids, vitamins, minerals, and antioxidants. Because of their versatility, egg, soy, and dairy products can be used across all menu categories to create flavorful menu items.

Match each egg-related characteristic or process with its description.

_______________ **1.** Yellow portion of an egg

_______________ **2.** Also known as albumen

_______________ **3.** Best used in recipes that call for raw or undercooked eggs

_______________ **4.** Egg product made from pasteurized eggs that have had the yolks removed and vegetable gum, coloring, and flavoring added

_______________ **5.** Yellow-colored liquid composed of soy, vegetable gums, and starches derived from corn or flour

_______________ **6.** Bacteria found in eggs that have become contaminated from infected laying hens

_______________ **7.** A process in which a hot liquid is gradually added to eggs to prevent curdling

_______________ **8.** Twisted cordlike strands of egg white that anchor the yolk to the center of the egg white

_______________ **9.** The amount of protein contained in one egg

_______________ **10.** The amount of fat contained in the yolk of one egg

A. Egg white
B. Eggless egg substitute
C. Egg yolk
D. Pasteurized eggs
E. Egg substitute
F. 5 g
G. Chalazae
H. 6 g
I. Salmonella enteritidis
J. Tempering

Match each soy product with its description.

_______________ **11.** Meatless product made from soybean by-products that is formed into the familiar shape of a processed meat product

_______________ **12.** Green soybeans

_______________ **13.** Asian sauce made from fermented soybeans, palm sugar, and usually star anise and garlic

_______________ **14.** Asian flavoring made from fermented soybeans, starch such as barley or rice, salt, and water

_______________ **15.** Flat, dense cake made from fermented soybeans

_______________ **16.** Curd of ground, cooked soybeans that has been pressed into a cake

_______________ **17.** Granular, meatless protein product made from soy flour

_______________ **18.** Protein skin that forms when soy milk is heated

_______________ **19.** Asian sauce made from fermented soybeans that has a distinct aroma and cheese-like flavor

_______________ **20.** Soy sauce made with little or no wheat; sauce is dark, thick, and mild in taste

_______________ **21.** Ground soybean pulp left over from the production of soy milk

_______________ **22.** Asian sauce made from fermented soybeans, wheat, salt, and water

A. Tofu
B. Soy sauce
C. Miso
D. Meat analog
E. Texturized soy protein
F. Yuba
G. Natto
H. Edamame
I. Okara
J. Tamari
K. Kecap
L. Tempeh

Match each liquid or semisolid dairy product with its description.

_______________ **23.** Semisolid dairy product that has been fermented by adding bacterial cultures to milk

_______________ **24.** Liquid dairy product produced by the mammary glands of cows, goats, sheep, or water buffalo

_______________ **25.** Fermented liquid dairy product made by adding bacterial cultures to low-fat or skim milk to create a thickened texture and tangy flavor

_______________ **26.** Semisolid dairy product that has been fermented by adding strains of active bacteria to cream

_______________ **27.** Semisolid dairy product that has been fermented by adding strains of active bacteria and yeast to milk

_______________ **28.** Liquid dairy product made from the butterfat that separates from non-homogenized milk

A. Buttermilk
B. Yogurt
C. Kefir
D. Cream
E. Milk
F. Sour cream

Match each type of cheese with its classification.

________________ **29.** Neufchâtel

________________ **30.** Swiss

________________ **31.** Gorgonzola

________________ **32.** Muenster

________________ **33.** Gruyère

________________ **34.** Monterey Jack

________________ **35.** Brie

________________ **36.** Stilton

________________ **37.** Asiago

________________ **38.** Havarti

________________ **39.** Ricotta

________________ **40.** Camembert

________________ **41.** Romano

________________ **42.** Brick

________________ **43.** Roquefort

________________ **44.** Mozzarella

________________ **45.** Cheddar

A. Fresh cheese
B. Soft cheese
C. Dry-rind cheese
D. Washed-rind cheese
E. Blue-veined cheese
F. Hard cheese
G. Grating cheese

Activity 9-2: Egg, Soy, and Dairy Product Menu Descriptions

Menu descriptions should be clear and appealing while communicating the main ingredients of a dish and the method of preparation. Effective menu descriptions promote clarity, add appeal, and increase the likelihood that an item will be ordered.

Rewrite the following ineffective menu descriptions to make them more appealing and complete.

1. Frittata

2. Egg Salad Sandwich

3. Tofu Stir-Fry

4. Yogurt Parfait

5. Grilled Cheese

Menu Descriptions Rubric					
Description Criteria	**Frittata**	**Egg Salad Sandwich**	**Tofu Stir-Fry**	**Yogurt Parfait**	**Grilled Cheese**
Accurate representation	1	1	1	1	1
Promotes clarity	1	1	1	1	1
Appealing portrayal	1	1	1	1	1
States unique characteristics	1	1	1	1	1
Conveys preparation techniques	1	1	1	1	1

Scoring: A point is earned for each menu item description that meets the criteria indicated in the rubric for a possible total score of 25 points.

Activity 9-3: Breakfast Sandwich Recipe Modification

The purpose of the culinary nutrition recipe-modification process is to increase the nutrient density of a menu item as well as to accommodate guests with dietary considerations such as food allergies. The appearance, texture, and flavor of menu items should also be considered when modifying recipes.

Modify the following breakfast sandwich recipe to reduce its calories, fat, and sodium, and to increase its dietary fiber, unsaturated fats, vitamins, and minerals.

Yield: 1 serving, 1 sandwich

Ingredients

2 tbsp	butter, divided
1 ea	croissant
2 ea	whole eggs
½ tsp	salt
2 oz	ham, sliced
2 oz	Cheddar cheese, sliced

Preparation

1. Add 1 tbsp of butter to a heated griddle.
2. Slice croissant in half lengthwise and place on the griddle to toast. Set aside.
3. Break eggs into a nonreactive bowl. Add salt and beat slightly with a whisk.
4. Add the remaining 1 tbsp of butter to the griddle.
5. Pour eggs onto the griddle and cook until eggs are set.
6. Place eggs on the bottom half of the croissant and top with ham and cheese.
7. Place the top half of the croissant on the sandwich and serve.

Culinary Nutrition Recipe-Modification Process

Step 1: Evaluate original recipe for sensory and nutritional qualities.

Step 2: Establish goals for recipe modifications.

Step 3: Identify modifications or substitutions.

Step 4: Determine functions of identified modifications or substitutions.

Step 5: Select appropriate modifications or substitutions.

Step 6: Test modified recipe to evaluate sensory and nutritional qualities.

Activity 9-4: Dairy Product Lab

Dairy products add diverse textures and flavors to menu items. Dairy products such as milk and yogurt are complete-protein foods that are rich in calcium. However, other dairy products, such as cream, are high in fat and low in calcium.

Objective: To identify milks, creams, and yogurts based on texture and flavor.

Materials and Equipment

- Marker
- 20 tasting cups
- Measuring cup
- Milks and creams
  - whole milk
  - 2% milk
  - nonfat milk
  - evaporated milk
  - half-and-half
  - heavy whipping cream
- Plain yogurts
  - regular
  - Greek
  - low-fat
  - nonfat
- 8 tasting spoons

Allergen Alert

The materials used in this activity may contain one or more food allergens.

Procedure

Milks and creams

1. Label the bottoms of six tasting cups with the name of a milk or cream, one milk or cream per label.
2. Place ¼ cup of each milk and cream into the correctly labeled cup.
3. Place a blindfold on one partner and have that partner identify each milk and cream based on its texture and flavor.
4. Repeat steps 1–3 with the other lab partner wearing the blindfold.

Yogurts

5. Label the bottoms of four tasting cups with the name of a yogurt, one yogurt per label.
6. Place ¼ cup of each yogurt into the correctly labeled cup.
7. Place a blindfold on one partner and have that partner identify each yogurt based on its texture and flavor.
8. Repeat steps 5–7 with the other lab partner wearing the blindfold.

Scoring: A point is earned for each milk, cream, and yogurt correctly identified based on its texture and flavor for a possible total score of 10 points.

Activity 9-5: Osteoporosis Research

Osteoporosis is a condition characterized by a loss of bone density that results in porous, brittle bones that are susceptible to breakage. Research indicates that diet and exercise play a role in producing strong, healthy bones.

Access an article from a medical-based website, a nutrition journal, and a company that sells supplements. Prepare a three- to four-page report and an oral presentation that address the following criteria.

1. Describe dietary recommendations for individuals with osteoporosis, including foods that should be added to the diet and foods that should be restricted.
2. Describe five foods other than dairy products that are good sources of calcium.
3. Identify the types of physical activities found to maintain and/or strengthen bone density and explain why these activities are beneficial.
4. Explain the role of supplements in the prevention and treatment of osteoporosis.
5. Plan a breakfast, a lunch, and a dinner menu (including beverages) that meets USDA calcium recommendations for your age and gender.

Nutrition Project Rubric

Content	
Written report	5
Cover page	5
Introduction	5
Quality content	20
Summary	5
Format	
Typed, double spaced	5
Correct spelling	5
Correct grammar	5
Correct punctuation	5
Bibliography	5
Oral Presentation	
Introduction	5
Quality content	10
Summary	5
Delivery	5
Visual aid(s)	10
Total	100

Poultry & Meats on the Menu

Chapter 10

REVIEW

Name ________________________________ **Date** ____________________

True-False

T F **1.** Slicing meats across the grain helps maintain tenderness.

T F **2.** Purchasing whole birds is more economical than purchasing fabricated pieces.

T F **3.** Meat is typically the least expensive item on the plate.

T F **4.** Leg and thigh flesh of poultry is always white.

T F **5.** Fanning meat on the plate gives the impression of a larger portion.

T F **6.** It is important to let cooked poultry rest so juices redistribute and keep the bird moist.

T F **7.** Cattle, hogs, and sheep provide most of the meat used in the professional kitchen.

T F **8.** Grain-fed beef contains less marbling than grass-fed beef.

T F **9.** Consuming the skin of poultry adds saturated fat and calories to a diet.

T F **10.** The cooking technique used to prepare poultry depends on the age and tenderness of the bird.

T F **11.** Ratite flesh is light pink and has a similar appearance to pork after it is cooked.

T F **12.** Grilling meats to the well-done stage is not recommended.

T F **13.** Dry rubs are rubbed onto the meat prior to cooking.

T F **14.** Farm-raised game birds include pheasants, quails, grouses, partridges, and wild turkeys.

T F **15.** The amount of marbling guarantees the flavor and tenderness of meat.

Multiple Choice

______________ **1.** A ___ cut is a ready-to-cook cut of meat that is packaged to certain size and weight specifications.
A. fabricated
B. primal
C. convenience
D. whole

______________ **2.** The more a muscle is used, the ___ the muscle becomes.
A. lighter and less flavorful
B. lighter and more flavorful
C. darker and less flavorful
D. darker and more flavorful

______________ **3.** Grass-fed bison meat contains less ___ than beef.
A. unsaturated fat
B. saturated fat
C. iron
D. protein

______________ **4.** ___ cuts of meat require longer cooking times and should be roasted at lower temperatures to prevent excessive shrinkage.
A. Lean
B. Tender
C. Larger
D. Smaller

______________ **5.** Even though ratites are classified as poultry, the flesh should be cooked no higher than ___°F since it will develop a metallic flavor at higher temperatures.
A. 125
B. 135
C. 145
D. 155

______________ **6.** ___ is the process of aging meat in vacuum-sealed plastic.
A. Wet aging
B. Dry aging
C. Curing
D. Rubbing

______________ **7.** Only ___ geese are used in foodservice operations.
A. young
B. mature
C. male
D. female

_______________ **8.** Chickens are classified by ___.
A. age only
B. gender only
C. age and sometimes gender
D. gender and weight

_______________ **9.** ___ is freshly ground or chopped raw meat or seafood.
A. Ceviche
B. Sashimi
C. Tartare
D. Carpaccio

_______________ **10.** The majority of fat in poultry is located just under the skin, near the tail, and on the ___.
A. giblets
B. wings
C. breasts
D. abdomen

_______________ **11.** All ground meats must be cooked to an internal temperature of ___°F.
A. 135
B. 140
C. 155
D. 160

_______________ **12.** Turkeys are classified by ___.
A. gender
B. age
C. weight
D. fat content

_______________ **13.** Because it has a similar texture and flavor to beef, ___ can be used in any recipe that calls for beef.
A. pork
B. lamb
C. rabbit
D. bison

_______________ **14.** Approximately one-third of ___ is sold fresh and two thirds is cured and smoked.
A. pork
B. lamb
C. veal
D. venison

_______ **15.** Chicken classifications include Cornish game hens, broilers/fryers, roasters, capons, and ___.
A. yearlings
B. grouses
C. steamers
D. stewers

_______ **16.** A duck yields ___ as much flesh as a chicken of the same size.
A. a quarter
B. half
C. twice
D. three times

_______ **17.** ___ is the process of cooking food by skewering it and suspending and rotating it above or next to a heat source.
A. Spit-roasting
B. Broiling
C. Barbequing
D. Smoking

_______ **18.** According to the USDA, common game meats include venison, rabbit, and ___.
A. bison
B. lamb
C. goat
D. pork

_______ **19.** For safe consumption, beef and veal should be cooked to an internal temperature of ___°F for at least 15 seconds and rested for 3 minutes.
A. 125
B. 135
C. 145
D. 155

_______ **20.** Chicken cacciatore and coq au vin are examples of ___ chicken dishes.
A. grilled
B. smoked
C. stewed
D. fried

_______ **21.** A young goose is commonly roasted at ___ temperatures to aid in rendering some of the fat from the skin and the flesh.
A. very low
B. low to medium
C. medium
D. very high

__________ **22.** ___ is trimmed from meat prior to cooking because it is inedible and removing it helps prevent meat from curling as it cooks.
A. Silverskin
B. Collagen
C. Marbling
D. Fat cap

__________ **23.** Pork is meat from slaughtered hogs that are ___ old.
A. less than a year
B. 1–2 years
C. 2–3 years
D. more than 3 years

__________ **24.** Marinades typically contain oil and a(n) ___ ingredient, which acts as a tenderizer in addition to providing flavor.
A. alkaline
B. acidic
C. sweet
D. umami

__________ **25.** Offals can include the heart, brain, kidney, and ___.
A. ribs
B. liver
C. shoulder
D. tenderloin

__________ **26.** Foie gras is the fattened liver of a duck or ___.
A. chicken
B. turkey
C. goose
D. pheasant

__________ **27.** ___ uses slower, combination cooking techniques that are effective on either tender or tougher cuts of meat.
A. Grilling
B. Barbequing
C. Stir-frying
D. Braising

__________ **28.** Ostriches and ___ are examples of ratites.
A. emus
B. grouses
C. capons
D. partridges

______________ **29.** The tenderloin is a ___ cut of meat.
- A. lean and tender
- B. lean and tough
- C. fatty and tender
- D. fatty and tough

______________ **30.** The fan, taken from a muscle in the ___, is the most tender cut of a ratite.
- A. wing
- B. breast
- C. thigh
- D. back

Completion

______________ **1.** ___ is the collective term for various kinds of birds that are raised for human consumption.

______________ **2.** ___ is the name for the grouping of the neck, gizzard, liver, and heart of a bird.

______________ **3.** ___ is the process of using salt and sodium nitrite alone or with flavorings or sugar to preserve a food item.

______________ **4.** A(n) ___ cut is a large cut of meat from a whole or a partial carcass.

______________ **5.** ___ is the process of using a brush or a ladle to place pan drippings or sauces over an item during the cooking process to help retain moisture and enhance flavor.

______________ **6.** A(n) ___ is the edible part of an animal that is not part of a primal cut.

______________ **7.** A(n) ___ is a substance added to a food to maintain or improve nutritional quality, maintain quality and freshness, aid in processing or preparation, or make food more appealing.

______________ **8.** A duck or goose that is bred to produce foie gras is fed a rich diet until the ___ becomes almost solid fat.

______________ **9.** ___ is the meat from calves, which are young cattle.

______________ **10.** Pork should be cooked to an internal temperature of ___°F for at least 15 seconds and rested for 3 minutes.

______________ **11.** ___ is a soft, white connective tissue that breaks down into gelatin when heated.

______________ **12.** ___ is meat from slaughtered sheep that are less than a year old.

______________ **13.** ___ is a dry-heat cooking technique in which food is slowly cooked over smoldering hardwoods in a vented enclosure.

______________ **14.** ___ is a tough, rubbery, silver-white connective tissue that does not break down when heated.

______________ **15.** A(n) ___ game bird is a game bird that is raised for legal sale.

Critical Thinking

Use complete sentences to respond to the following.

1. Choose a dry-heat, moist-heat, and combination cooking technique and describe how it can be used to prepare poultry.

2. Explain how poultry can be used throughout the menu mix.

3. Compare the ways in which poultry is commonly promoted at quick service and fast casual venues, casual dining and institutional venues, beverage houses and bar venues, and fine dining and special event venues.

4. Compare the types and cuts of meat commonly used for grilling, broiling, smoking, barbequing, roasting, sautéing, pan- and deep-frying, braising, and stewing.

5. Explain how aging, curing, rubs, and marinades are used to develop flavor and/or tenderness in meat.

6. Explain how meats can be effectively used in appetizers, soups, sandwiches, salads, and entrées.

Poultry & Meats on the Menu

Chapter 10

ACTIVITIES

Name ______________________________ **Date** ____________________

Activity 10-1: Nutrient Density of Poultry and Meat Products

The USDA National Nutrient Database for Standard Reference provides nutrient information for more than 8000 foods. Serving sizes can be modified to reflect dietary or recipe considerations.

Access the USDA National Nutrient Database for Standard Reference website at http://ndb.nal.usda.gov/ to complete the following chart.

	Product	Calories	Fat (g)	Sodium (mg)	Protein (g)
1.	Beef, rib, eye, small end, separable lean and fat, trimmed to 0″, choice, raw (3 oz)				
2.	Beef tenderloin, prime, broiled (3 oz)				
3.	Bratwurst, pork, cooked (85 g)				
4.	Burrito with beef, fast food (2)				
5.	Chicken, breaded and fried, light meat, fast food (2 pieces)				
6.	Chicken breast, oven-roasted roll (2 oz)				
7.	Chicken, canned, with broth (2.5 oz)				
8.	Deli ham, sliced, regular (56 g)				
9.	Frankfurter, beef, unheated (50 g)				
10.	Pizza, sausage, regular crust, fast food (1 slice)				
11.	Pork chop, center loin, bone-in, separable lean and fat, broiled (3 oz)				

Activity 10-2: Poultry and Meat Menu Descriptions

Menu descriptions should be clear and appealing while communicating the main ingredients of a dish and the method of preparation. Effective menu descriptions promote clarity, add appeal, and increase the likelihood that the item will be ordered.

Rewrite the following ineffective menu descriptions to make them more appealing and complete.

1. Chicken Satay

2. Beef Stew

3. Veal Piccata

4. Pulled Pork Sandwich

5. Spice-Rubbed Lamb Chops

Menu Descriptions Rubric					
Description Criteria	Chicken Satay	Beef Stew	Veal Piccata	Pulled Pork Sandwich	Spice-Rubbed Lamb Chops
Accurate representation	1	1	1	1	1
Promotes clarity	1	1	1	1	1
Appealing portrayal	1	1	1	1	1
States unique characteristics	1	1	1	1	1
Conveys preparation techniques	1	1	1	1	1

Scoring: A point is earned for each menu item description that meets the criteria indicated in the rubric for a possible total score of 25 points.

Activity 10-3: Chile Con Carne Recipe Modification

The purpose of the culinary nutrition recipe-modification process is to increase the nutrient density of a menu item as well as to accommodate guests with dietary considerations such as food allergies. The appearance, texture, and flavor of menu items should also be considered when modifying recipes.

Modify the following chili recipe to reduce its calories, fat, and sodium and to create a poultry menu item.

Yield: 4 servings, 1¼ cup each

Ingredients

1 tbsp	vegetable oil
1 lb	ground beef
1 c	onions, chopped
1 c	green peppers, chopped
1 ea	tomatoes, 28 oz canned
1 ea	tomato sauce, 8 oz canned
2 tsp	chili powder
1 tsp	salt
⅛ tsp	cayenne pepper
⅛ tsp	paprika
1 ea	kidney beans, 15½ oz canned, drained
1 c	Cheddar cheese, shredded
1 c	sour cream

Preparation

1. Heat oil in a sauté pan over medium heat. Add the beef, onions, and green peppers and cook until the beef is browned.
2. Add the canned tomatoes, tomato sauce, chili powder, salt, cayenne pepper, and paprika. Stir to combine.
3. Bring to a boil and then reduce heat, cover, and simmer for 2 hours.
4. After 2 hours, remove the lid and simmer for 45 minutes.
5. Add the beans and stir to combine.
6. Portion the chili into four bowls and garnish each bowl with ¼ cup Cheddar cheese and ¼ cup sour cream.

Culinary Nutrition Recipe-Modification Process

Step 1: Evaluate original recipe for sensory and nutritional qualities.

Step 2: Establish goals for recipe modifications.

Step 3: Identify modifications or substitutions.

Step 4: Determine functions of identified modifications or substitutions.

Step 5: Select appropriate modifications or substitutions.

Step 6: Test modified recipe to evaluate sensory and nutritional qualities.

Activity 10-4: Poultry Lab

Variations in the appearance and flavor of different types of poultry can be subtle or well-defined. Often, one type of poultry can be substituted for another type with only a slight change in appearance and flavor.

Objective: To determine the variations among cooked chicken, turkey, and duck in terms of appearance and flavor.

Materials and Equipment

- Nonstick cooking spray
- Sheet pan
- Chicken breast
- Turkey breast
- Duck breast
- Instant read thermometer
- 3 plates
- Knife
- Fork

Procedure

1. Place the chicken, turkey, and duck breasts on a sheet pan that has been sprayed with nonstick cooking spray.
2. Roast at 350°F until each breast reaches a minimum internal temperature of 165°F.
3. Remove the breasts from the oven and allow them to rest for approximately 5–10 minutes.
4. Evaluate each type of poultry in terms of appearance and flavor.

Respond to the following.

1. Which type of poultry has the darkest color?

2. Which type of poultry has the lightest color?

3. Which type of poultry has the strongest flavor?

4. Which type of poultry has the mildest flavor?

Activity 10-5: Meats Lab

The appearance and flavor vary among different types of meat. Understanding these variations means the meat can be prepared using complementary ingredients that enhance its flavor.

Objective: To determine the variations among cooked veal, pork, and lamb in terms of appearance and flavor.

Materials and Equipment

- Vegetable oil, 1 oz (for grill)
- Veal chop
- Pork chop
- Lamb chop
- Tongs
- Instant read thermometer
- 3 plates
- Knife
- Fork

Procedure

1. Place the veal, pork, and lamb chops on a heated and oiled grill.
2. Grill the chops on one side and then use the tongs to flip them.
3. Continue cooking until each chop reaches a minimum internal temperature of 145°F.
4. Remove the chops from the grill and allow them to rest for approximately 5–10 minutes.
5. Evaluate each type of meat in terms of appearance and flavor.

Respond to the following.

1. Which type of meat has the darkest color?

2. Which type of meat has the lightest color?

3. Which type of meat has the strongest flavor?

4. Which type of meat has the mildest flavor?

Activity 10-6: Livestock and Sustainable Agriculture Research

Livestock such as poultry, cattle, hogs, and sheep are often raised using sustainable agriculture practices. Sustainable agriculture protects the welfare of livestock by implementing practices that focus on the health and well-being of the animals.

Access the National Agricultural Library website at http://afsic.nal.usda.gov/sustainability-agriculture-0. Prepare a three- to four-page report and an oral presentation that address the following criteria.

1. Explain the meaning of sustainable agriculture.
2. Describe sustainable practices that are used with poultry, cattle, hogs, and sheep.
3. Explain how sustainable practices impact the animals in terms of nutrition.
4. Describe how sustainable practices impact food costs.
5. Explain how this research will affect your selection of poultry and meat products in the future.

Nutrition Project Rubric

Content	
Written report	5
Cover page	5
Introduction	5
Quality content	20
Summary	5
Format	
Typed, double spaced	5
Correct spelling	5
Correct grammar	5
Correct punctuation	5
Bibliography	5
Oral Presentation	
Introduction	5
Quality content	10
Summary	5
Delivery	5
Visual aid(s)	10
Total	100

Fish & Shellfish on the Menu

Chapter 11

REVIEW

Name ______________________ **Date** ______________

True-False

T F **1.** Live bivalves that close when tapped should be discarded.

T F **2.** Farm-raised fish are leaner than wild-caught fish of the same varieties.

T F **3.** Color is a good indicator of the doneness of shellfish.

T F **4.** The muscle tissues of fish are arranged in layers that flake when the connective tissue gelatinizes during the cooking process.

T F **5.** Lobster flesh cannot tolerate the intense heat of a broiler.

T F **6.** The flavor profile of shellfish is often enhanced with an acidic ingredient.

T F **7.** Grilling is recommended for lean, delicate fish.

T F **8.** Fish should only be marinated for a short period of time.

T F **9.** Freshwater fish typically have more flavor than saltwater fish.

T F **10.** Shellfish are typically added to soups at the beginning of the cooking process.

T F **11.** Only fatty fish are suitable for sautéing.

T F **12.** Fish is a delicate food that is ideal for poaching.

T F **13.** When shellfish are sold fresh, they are often still alive.

T F **14.** Seaweed should not be added to steamed shellfish because it will dry the shellfish out.

T F **15.** Mentioning the point of origin for shellfish in a menu description can be a powerful selling tool.

Multiple Choice

______________ **1.** Canned fish packed in water instead of oil contains ___.
A. fewer calories
B. more calories
C. less trans fat
D. more trans fat

______________ **2.** Catfish, trout, herring, salmon, mackerel, and ___ are considered fatty fish.
A. tilapia
B. tuna
C. Northern pike
D. cod

______________ **3.** A ___ is a mollusk that has a single solid shell and a single foot.
A. crustacean
B. cephalopod
C. univalve
D. bivalve

______________ **4.** There are two classifications of shellfish: ___.
A. crustaceans and mollusks
B. crustaceans and exoskeletons
C. mollusks and cephalopods
D. mollusks and univalves

______________ **5.** A ___ fish is a fish that has had only the viscera removed.
A. drawn
B. dressed
C. pan-dressed
D. butterflied

______________ **6.** ___ is imitation shellfish made from fish that is colored, flavored, and shaped to resemble a particular type of shellfish.
A. Caviar
B. Ceviche
C. Surimi
D. Sashimi

______________ **7.** Only ___ roe can be labeled as caviar.
A. sturgeon
B. salmon
C. perch
D. whitefish

_______________ **8.** ___ are crustaceans.
A. Prawns
B. Scallops
C. Abalone
D. Squid

_______________ **9.** Crustaceans always have a(n) ___.
A. beard
B. cuttlebone
C. external shell
D. siphon

_______________ **10.** Flounder, halibut, and ___ are flatfish.
A. cod
B. perch
C. sole
D. salmon

_______________ **11.** Overcooking fish causes the ___ in the fish to shrink and squeeze out water, which results in a dry finished product.
A. proteins
B. carbohydrates
C. fats
D. vitamins

_______________ **12.** Cephalopods include squids, octopuses, and ___.
A. prawns
B. scallops
C. cuttlefish
D. skates

_______________ **13.** Bivalves include clams, cockles, mussels, oysters, and ___.
A. crayfish
B. conchs
C. abalones
D. scallops

_______________ **14.** ___ shellfish creates an unappealing texture.
A. Boiling
B. Simmering
C. Poaching
D. Steaming

_______________ **15.** Crustaceans, such as shrimp and lobster, and cephalopods, such as squid, are higher in ___ than other seafood.
A. saturated fat
B. cholesterol
C. sugar
D. dietary fiber

_______________ **16.** Fresh fish is best stored between 30°F and 34°F and used within ___ days.
A. 1–2
B. 2–3
C. 4–6
D. 6–8

_______________ **17.** Mollusks contain a ___, which is a tubular organ that is used to draw in or eject fluids.
A. cuttlebone
B. siphon
C. beard
D. foot

_______________ **18.** ___ fish are fish that are not endangered or overharvested.
A. Saltwater
B. Freshwater
C. Wild-caught
D. Sustainable

_______________ **19.** Sharks, skates, and stingrays are ___.
A. flatfish
B. cartilaginous fish
C. crustaceans
D. cephalopods

_______________ **20.** The only part of a lobster that is sold frozen is the ___.
A. head
B. thorax
C. claw
D. tail

Completion

_______________ **1.** A(n) ___ is a shellfish that has a hard, segmented shell protecting its soft flesh and lacks an internal bone structure.

_______________ **2.** A(n) ___ is a mollusk that does not have an external shell.

_______________ **3.** The ___ attempts to standardize fish nomenclature and provides the market and common names for both imported and domestic fish.

_______________ **4.** A(n) ___ is a presentation of a variety of raw and steamed seafood presented and served on a bed of ice.

_______________ **5.** A(n) ___ fish is a dressed fish that has had its head removed.

_______________ **6.** A(n) ___ is a thin, wide fish with both eyes located on one side of the head and a backbone that runs from head to tail through the lateral line of its body.

_______________ **7.** A fish ___ is a lengthwise piece of flesh cut away from the backbone of a fish.

_______________ **8.** The dark ___ from cephalopods, such as cuttlefish, can add color to pastas, sauces, and risottos and result in intriguing presentations.

_______________ **9.** A(n) ___ is a mollusk that has a top shell and a bottom shell connected by a central hinge.

_______________ **10.** A(n) ___ is a fish with a cylindrical body, an eye located on each side of the head, and a backbone that runs from head to tail in the center of its body.

_______________ **11.** ___ is a vinegar-seasoned rice dish garnished with raw fish, cooked seafood, eggs, or vegetables.

_______________ **12.** ___ is extremely thin slices of raw fish that have been marinated in lemon or lime juice.

_______________ **13.** ___ are grouped into univalves, bivalves, and cephalopods.

_______________ **14.** ___ is very thin slices of raw fish served with garnishes and sauces.

_______________ **15.** When a fish is wrapped in parchment paper and cooked, it is cooked "___."

Critical Thinking

Use complete sentences to respond to the following.

1. Compare lean and fatty fish in terms of nutrients, flavor, and habitat.

2. Describe the types of fish best prepared with dry-heat cooking techniques such as grilling, broiling, smoking, baking, sautéing, and frying.

3. Describe how beverage houses and bar venues promote shellfish on the menu.

4. Explain how to develop flavor in fish and shellfish dishes.

Fish & Shellfish on the Menu — Chapter 11 ACTIVITIES

Name ______________________________ **Date** ________________

Activity 11-1: Fish and Shellfish

Lean fish contain very little fat and typically have light-colored flesh and a mild flavor. Fatty fish contain more fat, are rich in omega-3 fatty acids, and typically have dark-colored flesh that is strong in flavor. Nutritionally, shellfish are similar to fish. Like fish, shellfish are also low in saturated fat. However, some shellfish do contain moderate amounts of cholesterol.

Identify each fish as either a lean fish or a fatty fish.

______________ **1.** Cod

______________ **2.** Perch

______________ **3.** Mackerel

______________ **4.** Herring

______________ **5.** Haddock

______________ **6.** Tilapia

______________ **7.** Trout

______________ **8.** Halibut

______________ **9.** Salmon

______________ **10.** Tuna

______________ **11.** Northern pike

______________ **12.** Catfish

Access the USDA National Nutrient Database for Standard Reference website at http://ndb.nal.usda.gov/ to determine the amount of cholesterol in the following shellfish.

	Raw Shellfish (3 oz)	Cholesterol (mg)
13.	Shrimp	
14.	Northern lobsters	
15.	Dungeness crabs	
16.	Clams	
17.	Blue mussels	
18.	Pacific oysters	
19.	Scallops	
20.	Squids	

Activity 11-2: Fish and Shellfish Menu Descriptions

Menu descriptions should be clear and appealing while communicating the main ingredients of a dish and the method of preparation. Effective menu descriptions promote clarity, add appeal, and increase the likelihood that the item will be ordered.

Rewrite the following ineffective menu descriptions to make them more appealing and complete.

1. Shrimp Cocktail

2. Ceviche

3. Bouillabaisse

4. Salmon en Papillote

5. Sole Almondine

Menu Descriptions Rubric					
Description Criteria	**Shrimp Cocktail**	**Ceviche**	**Bouillabaisse**	**Salmon en Papillote**	**Sole Almondine**
Accurate representation	1	1	1	1	1
Promotes clarity	1	1	1	1	1
Appealing portrayal	1	1	1	1	1
States unique characteristics	1	1	1	1	1
Conveys preparation techniques	1	1	1	1	1

Scoring: A point is earned for each menu item description that meets the criteria indicated in the rubric for a possible total score of 25 points.

Activity 11-3: Fish and Chips Recipe Modification

The purpose of the culinary nutrition recipe-modification process is to increase the nutrient density of a menu item as well as to accommodate guests with dietary considerations such as food allergies. The appearance, texture, and flavor of menu items should also be considered when modifying recipes.

Modify the following fish and chips recipe to reduce its calories, fat, and sodium, and to increase its dietary fiber, vitamins, and minerals.

Yield: 4 servings

Ingredients

3 c	vegetable oil
1 c	water
1 c	all-purpose flour
1 tsp	salt
½ tsp	pepper
1½ lb	cod fillets, cut lengthwise into 1½ inch strips
2 lb	russet potatoes, peeled, cut lengthwise into ½ inch strips
TT	kosher salt

Preparation

Fish

1. Heat ½ inch of oil in a heavy skillet to 360°F.
2. Whisk together water, flour, salt, and pepper until smooth.
3. Add fish to the water-flour batter and stir to coat.
4. Carefully place the fish in the hot oil and fry until golden (approximately 3 minutes).
5. Place fish on paper towels and season with salt and pepper.

Chips

6. Carefully place the potatoes in the hot oil and fry until golden (approximately 7 minutes).
7. Place potatoes on paper towels and season with kosher salt to taste.

Culinary Nutrition Recipe-Modification Process

Step 1: Evaluate original recipe for sensory and nutritional qualities.

Step 2: Establish goals for recipe modifications.

Step 3: Identify modifications or substitutions.

Step 4: Determine functions of identified modifications or substitutions.

Step 5: Select appropriate modifications or substitutions.

Step 6: Test modified recipe to evaluate sensory and nutritional qualities.

Activity 11-4: Ceviche Lab

Ceviche is a dish where fish is "cooked" using the acidity of citrus juices. The fish used to make ceviche must be fresh and have a firm texture. Fresh chopped shellfish such as shrimp, scallops, or squid may be substituted for fish.

Objective: To evaluate the texture of fish that is marinated for different lengths of time.

Materials and Equipment

- 3 nonreactive bowls
- Measuring cup
- Measuring spoons
- Mixing spoon
- 2 plates
- 2 forks
- Ceviche ingredients:

8 oz	fresh fish fillet (e.g., grouper, red snapper), small dice
¼ c	lime juice
1 tbsp	orange juice
1 tbsp	pineapple juice
1 tbsp	olive oil
1 tsp	jalapeno, seeded and minced
1 ea	garlic clove, minced
¼ tsp	salt
¼ tsp	pepper

Allergen Alert

The materials used in this activity may contain one or more food allergens.

Procedure

1. Add lime juice, orange juice, pineapple juice, olive oil, jalapeno, garlic, salt, and pepper in one of the nonreactive bowls. Stir to combine.
2. Add diced fish and gently toss until the fish is well-coated with the marinade.
3. Place half of the ceviche in the second nonreactive bowl. Cover with plastic wrap and refrigerate for 4 hours.
4. Place the remaining half of the ceviche in the third nonreactive bowl. Cover with plastic wrap and refrigerate 8 hours (or overnight).
5. Taste and evaluate the texture of the ceviche marinated for 4 hours.
6. Taste and evaluate the texture of the ceviche marinated for 8 hours (or overnight).

Respond to the following.

1. Compare the texture of the ceviche that was marinated for 4 hours with the ceviche that was marinated for 8 hours (or overnight).

2. Explain the impact the acidic marinade has on proteins and how this affects the texture of the ceviche in terms of marinating time.

Activity 11-5: Sustainable Seafood Research

Sustainable seafood includes fish and shellfish that are not endangered or overharvested. Several organizations monitor fishing waters and harvesting methods in order to provide consumers with lists of the most sustainable seafood to use as well as seafood to avoid.

Access the following websites:

Monterey Bay Aquarium Seafood Watch®: www.montereybayaquarium.org/cr/seafoodwatch.aspx
Environmental Defense Fund Seafood Selector: http://seafood.edf.org/

Prepare a three- to four-page report and an oral presentation that address the following criteria.

1. Describe three issues that threaten the wild-caught seafood supply.
2. Describe three issues that threaten the aquafarmed seafood supply.
3. Identify three common contaminants found in fishing waters.
4. Name the sources of these contaminants.
5. Describe the impact of these contaminants on ocean wildlife.
6. Identify seafood that is likely to contain these contaminants.
7. Explain why some seafood is more likely to contain these contaminants.
8. Identify which population groups should limit their intake of specific types of seafood and explain why.
9. Explain how this research will affect your selection of seafood in the future.

Nutrition Project Rubric

Content	
Written report	5
Cover page	5
Introduction	5
Quality content	20
Summary	5
Format	
Typed, double spaced	5
Correct spelling	5
Correct grammar	5
Correct punctuation	5
Bibliography	5
Oral Presentation	
Introduction	5
Quality content	10
Summary	5
Delivery	5
Visual aid(s)	10
Total	100

Vegetables & Legumes on the Menu

Chapter 12

REVIEW

Name ______________________________ **Date** ____________________

True-False

T F **1.** Because legumes are high in protein, they are ideal for vegetarian dishes.

T F **2.** Blanching vegetables destroys all nutrients and yields limited flavor.

T F **3.** White vegetables contain beta-carotene.

T F **4.** Unlike most nutrients, lycopene is better absorbed by the body when foods are cooked.

T F **5.** Legumes that are overcooked will become firm and rigid.

T F **6.** Canned legumes are typically preserved in a salt solution and should be rinsed off before being prepared.

T F **7.** When legumes are the featured item in vegetarian entrées, they should be served with whole grains to provide complete proteins.

T F **8.** Many varieties of dried legumes need to be soaked before being cooked.

T F **9.** Puréed black beans can be added to brownies as a fat replacement.

T F **10.** Legumes cannot be eaten raw.

T F **11.** Most fresh vegetables should be stored between 55°F and 60°F.

T F **12.** Drying vegetables can result in the loss of water-soluble vitamins.

T F **13.** Lentils cannot be cooked without being soaked first.

T F **14.** Frozen vegetables do not retain nutrients.

T F **15.** Vegetarian diets vary and can be classified by the types of foods consumed.

Multiple Choice

_______________ **1.** ___ have a higher protein content than other vegetables and therefore are often considered a protein source as well as a starch.
- A. Legumes
- B. Mushrooms
- C. Fruit-vegetables
- D. Root vegetables

_______________ **2.** ___ vegetables are often sautéed or roasted and used as building blocks of flavor as well as for their aromatic qualities.
- A. Stem
- B. Leaf
- C. Bulb
- D. Seed

_______________ **3.** A ___ is the oval or kidney-shaped edible seed of various plants in the legume family.
- A. bean
- B. pea
- C. lentil
- D. sprout

_______________ **4.** With the exception of ___, most legumes are low in fat.
- A. lima beans
- B. kidney beans
- C. soybeans
- D. lentils

_______________ **5.** Vegetables supply the body with nutrients such as ___ and an abundant amount of vitamins and minerals.
- A. saturated fat
- B. hydrogenated fat
- C. complete proteins
- D. dietary fiber

_______________ **6.** When simmering fibrous vegetables, starting them in ___ water allows the fibers in the vegetables to soften more evenly and create a consistent texture in the final product.
- A. cold
- B. hot
- C. simmering
- D. boiling

_______________ **7.** Examples of ___ include potatoes, sweet potatoes, yams, ocas, sunchokes, and water chestnuts.
- A. root vegetables
- B. bulb vegetables
- C. tubers
- D. fungi

_______________ **8.** Most ___ must be rehydrated before they can be cooked.
- A. edible pods
- B. split peas
- C. lentils
- D. pulses

_______________ **9.** A ___ is a very small pulse that has been split in half.
- A. lentil
- B. sprout
- C. pea
- D. bean

_______________ **10.** Tomatoes contain a high amount of ___, which has been found to reduce several types of cancer, especially prostate cancer.
- A. lycopene
- B. beta-carotene
- C. chlorophyll
- D. anthoxanthin

_______________ **11.** After dried legumes have been soaked, they should be ___.
- A. sprinkled with salt
- B. sprinkled with sugar
- C. rinsed with fresh, cold water
- D. rinsed with boiling water

_______________ **12.** One of the most effective cooking techniques used to caramelize vegetables and develop flavors is ___.
- A. simmering
- B. roasting
- C. blanching
- D. steaming

_______________ **13.** Sea vegetables have a naturally ___ taste because of the minerals they absorb from the ocean.
- A. bitter
- B. sweet
- C. sour
- D. salty

____________________ **14.** Examples of ___ vegetables include cauliflower, broccoli, squash blossoms, and artichokes.
- A. stem
- B. root
- C. flower
- D. fruit-

____________________ **15.** A ___ is an edible strand of a germinated bean or seed.
- A. pulse
- B. sprout
- C. tuber
- D. fungus

____________________ **16.** ___ vegetables lack more nutrients than other forms of vegetables because they are packaged in liquid, which facilitates the loss of water-soluble nutrients.
- A. Fresh
- B. Frozen
- C. Canned
- D. Dried

____________________ **17.** The classic combination of carrots, celery, and ___ known as mirepoix is often used to develop flavors in items such as stocks, broths, soups, and stews.
- A. onions
- B. garlic
- C. leeks
- D. turnips

____________________ **18.** ___ vegetables contain anthocyanins, which act as powerful antioxidants.
- A. White
- B. Green
- C. Yellowish-orange
- D. Bluish-purple

____________________ **19.** Leaf vegetables have a mild flavor when they are young, but their flavor becomes ___ with age.
- A. bitter
- B. sweet
- C. salty
- D. sour

____________________ **20.** Legumes are essential menu items for vegetarians because they contain ___.
- A. cholesterol
- B. starch
- C. nonessential amino acids
- D. essential amino acids

Completion

______________ **1.** A(n) ___ is a strongly flavored vegetable that grows underground and consists of a short stem base with one or more buds that are enclosed in overlapping membranes or leaves.

______________ **2.** A(n) ___ is the edible seed of a nonwoody plant and grows in multiples within a pod.

______________ **3.** ___ vegetables include carrots, parsnips, salsify, radishes, turnips, rutabagas, beets, celeriac, jicamas, lotus roots, and bamboo shoots.

______________ **4.** A(n) ___ is the fleshy, spore-bearing body of an edible fungus that grows above the ground.

______________ **5.** A(n) ___ is the dried seed of a legume.

______________ **6.** Cabbages, bok choy, Brussels sprouts, lettuces, chicory, watercress, spinach, chard, kale, and mustard greens are classified as ___.

______________ **7.** A(n) ___ is a botanical fruit that is sold, prepared, and served as a vegetable.

______________ **8.** Green vegetables contain the pigment ___.

______________ **9.** Legumes such as beans, peas, pulses, and lentils are also classified as ___ vegetables.

______________ **10.** A(n) ___ is the round edible seed of various plants in the legume family.

______________ **11.** Common varieties of ___ include nori, kombu, arame, wakame, dulse, and hijiki.

______________ **12.** ___ vegetables include asparagus, celery, fennel, rhubarb, kohlrabi, and hearts of palm.

______________ **13.** ___ is the process of preserving foods in an airtight container to prevent spoilage.

______________ **14.** Orange and yellow vegetables, including butternut squash, carrots, pumpkin, rutabagas, and sweet corn, are colored by ___, which converts to vitamin A in the body.

______________ **15.** Some bean varieties, such as green beans, are called ___ because both the pod and the beans are edible.

Critical Thinking

Use complete sentences to respond to the following.

1. Describe the nutritional benefits of vegetables and legumes.

2. Describe the health benefits associated with specific color pigments found in vegetables.

3. Summarize how the market form of the legume purchased affects the preparation procedures.

Vegetables & Legumes on the Menu

Chapter 12

ACTIVITIES

Name ______________________________ **Date** ____________________

Activity 12-1: Sodium Content in Vegetables and Legumes

Vegetable selection, storage, and preparation vary depending on the form of the vegetables purchased. Vegetables can be purchased fresh, frozen, canned, or dried.

Access the USDA National Nutrient Database for Standard Reference website at http://ndb.nal.usda.gov/ to determine the amount of sodium, in milligrams, contained in a 100 g serving of each form of the type of vegetable listed.

Type	Fresh, Raw	Frozen, Unprepared	Canned, Drained	Dried
Beans, snap, green		*(all styles)*		N/A
Carrots				*(dehydrated)*
Corn, sweet, yellow		*(off the cob)*		N/A
Onions		*(whole)*	*(solids and liquids)*	*(dehydrated)*
Peas, green			*(solids and liquids)*	N/A
Peppers, sweet, red			*(solids and liquids)*	*(freeze-dried)*
Pinto beans	*(mature)*	*(immature)*		N/A
Potatoes	*(flesh and skin)*	*(whole)*		N/A
Spinach				N/A
Tomatoes, red, ripe		N/A	*(stewed)*	*(sun-dried)*

Activity 12-2: Vegetable and Legume Menu Descriptions

Menu descriptions should be clear and appealing while communicating the main ingredients of a dish and the method of preparation. Effective menu descriptions promote clarity, add appeal, and increase the likelihood that the item will be ordered.

Rewrite the following ineffective menu descriptions to make them more appealing and complete.

1. Crudités with Dip

2. Vegetable-Bean Soup

3. Veggie Burger

4. Heirloom Bean Salad

5. Broccoli Slaw

Menu Descriptions Rubric					
Description Criteria	**Crudités with Dip**	**Vegetable-Bean Soup**	**Veggie Burger**	**Heirloom Bean Salad**	**Broccoli Slaw**
Accurate representation	1	1	1	1	1
Promotes clarity	1	1	1	1	1
Appealing portrayal	1	1	1	1	1
States unique characteristics	1	1	1	1	1
Conveys preparation techniques	1	1	1	1	1

Scoring: A point is earned for each menu item description that meets the criteria indicated in the rubric for a possible total score of 25 points.

Activity 12-3: Vegetable Polonaise Recipe Modification

The purpose of the culinary nutrition recipe-modification process is to increase the nutrient density of a menu item as well as to accommodate guests with dietary considerations such as food allergies. The appearance, texture, and flavor of menu items should also be considered when modifying recipes.

Modify the following vegetable polonaise recipe to reduce its calories, saturated fat, and sodium; enhance its colors, textures, and flavors; and increase its dietary fiber, vitamins, and minerals.

Yield: 8 servings, ⅔ cup each

Ingredients

1 tbsp	salt
4 c	cauliflower florets
4 c	broccoli florets
2 tbsp	butter, melted
¾ c	bread crumbs
2 ea	whole, hard-cooked eggs, finely chopped
2 tbsp	parsley, minced
3 tbsp	heavy cream
1 tsp	kosher salt

Preparation

1. Bring 10 cups of water to a boil and add salt.
2. Add the cauliflower and broccoli and cook until al dente (tender-crisp, with some resistance to the bite).
3. Place the vegetables in an ice bath to stop the cooking process.
4. Remove the vegetables and allow them to air dry for approximately 5 minutes.
5. Melt butter in a medium skillet over medium-high heat.
6. Add bread crumbs, eggs, and parsley and stir to combine. Set aside.
7. Heat heavy cream in a large skillet over medium heat.
8. Add the vegetables and cook until heated through.
9. Transfer the vegetables to a serving bowl and season with kosher salt.
10. Top with the bread crumb mixture.

Culinary Nutrition Recipe-Modification Process

Step 1: Evaluate original recipe for sensory and nutritional qualities.

Step 2: Establish goals for recipe modifications.

Step 3: Identify modifications or substitutions.

Step 4: Determine functions of identified modifications or substitutions.

Step 5: Select appropriate modifications or substitutions.

Step 6: Test modified recipe to evaluate sensory and nutritional qualities.

Activity 12-4: Vegetable Reactions to Acids and Alkalis Lab

An acidic or an alkali ingredient can affect the color and texture of vegetables during cooking and alter the natural pigments in vegetables. Pigments include the carotenoids found in yellow and orange vegetables, the chlorophyll found in green vegetables, and the flavonoids found in purple, dark red, and white vegetables.

Objective: To determine the effect acid and alkali cooking solutions have on the color and texture of vegetables with different pigments.

Materials and Equipment

- Labels
- Marker
- 9 plates
- Cutting board
- Chef's knife
- 3 carrots
- 3 leaves red cabbage
- 15 green beans
- 3 medium sauce pans
- Water
- Measuring cup
- 1 cup white vinegar
- Tablespoon
- 3 tbsp baking soda
- Slotted spoon

Procedure

1. Label three plates "water," three plates "acid," and three plates "alkali."
2. Cut carrots into 1-inch rounds and divide into three equal piles.
3. Fill the three sauce pans halfway with water and bring to a boil.
4. Add ⅓ cup vinegar to one of the pans and 1 tbsp baking soda to another pan.
5. Place a third of the carrots into each of the three pans and boil for 1 minute.
6. Turn off the heat and allow the carrots to remain in the hot water for an additional minute.
7. Use the slotted spoon to remove the carrots from the pure water to the "water" plate, the carrots from the vinegar water to the "acid" plate, and the carrots from the baking soda water to the "alkali" plate.
8. Clean the sauce pans.
9. Repeat steps 3–8 for the red cabbage and then the green beans.
10. Evaluate each plate of vegetables in terms of color and texture.

Respond to the following.

1. Compare how the carotenoid pigment found in carrots reacts to a water, acid, and alkali solution in terms of color and texture.

2. Compare how the chlorophyll pigment found in green beans reacts to a water, acid, and alkali solution in terms of color and texture.

3. Compare how the flavonoid pigment found in red cabbage reacts to a water, acid, and alkali solution in terms of color and texture.

Activity 12-5: Cancer Research

Vegetables and legumes are abundant in antioxidants and phytonutrients, which are also known as phytochemicals. A diet plentiful in a variety of vegetables and legumes has many health benefits, such as reducing the risk of developing certain types of cancer.

Access the American Institute for Cancer Research at http://www.aicr.org/. Prepare a three- to four-page report and an oral presentation that address the following criteria.

1. Describe the correlation between diet and physical activity and the risk of developing cancer.
2. Identify a total of five vegetables and/or legumes that have been found to reduce the risk of developing cancer.
3. Name the key nutrients in these vegetables and legumes.
4. Describe the health benefits of phytonutrients.
5. Identify five phytonutrients found in vegetables and/or legumes.
6. Summarize the potential health benefits of these phytonutrients.
7. Plan a meal that is a good source of three phytonutrients.

Nutrition Project Rubric

Content	
Written report	5
Cover page	5
Introduction	5
Quality content	20
Summary	5
Format	
Typed, double spaced	5
Correct spelling	5
Correct grammar	5
Correct punctuation	5
Bibliography	5
Oral Presentation	
Introduction	5
Quality content	10
Summary	5
Delivery	5
Visual aid(s)	10
Total	100

Fruits, Nuts & Seeds on the Menu

Chapter 13

REVIEW

Name ______________________________ **Date** ____________________

True-False

T F **1.** Descriptions of fruits on the menu should highlight the name of the fruit and the preparation techniques.

T F **2.** Melons are harvested just before ripeness.

T F **3.** Nuts contain a moderate amount of cholesterol.

T F **4.** Nuts and seeds should only be added in the beginning stages of preparation.

T F **5.** Nut-based milks are lactose-free.

T F **6.** The vibrant colors and sweet flavors of fruits do not pair well with animal-based proteins.

T F **7.** Citrus fruits can enhance the saltiness of a dish.

T F **8.** Flaxseeds are one of the richest sources of omega-3 fatty acids.

T F **9.** Although many fruits can be poached, firmer fruits such as pears are often used for poaching.

T F **10.** Nuts and seeds do not easily absorb flavors.

T F **11.** The oil extracted from sunflower seeds cannot be used for cooking oil.

T F **12.** Frozen fruits are softer than fresh fruits because the cell membranes of the fruit rupture and release water during the freezing process.

T F **13.** Nuts and seeds may increase the protein content of a dish, but they are high in unhealthy fats and low in micronutrients.

T F **14.** Tropical fruits tend to be more recognizable than exotic fruits because they are more widely cultivated.

T F **15.** The sweetness of fruits is naturally compatible with the bitter flavors of many salad greens.

Multiple Choice

_______________ **1.** ___ fruits supply anthoxanthins, which have been found to contain health-promoting chemicals that lower blood pressure and cholesterol.
- A. White
- B. Red
- C. Green
- D. Purple

_______________ **2.** A ___ fruit is a result of crossbreeding two or more fruits of different species to obtain a completely new fruit.
- A. variety
- B. hybrid
- C. cultivated
- D. classified

_______________ **3.** ___ have the highest ratio of heart-healthy monounsaturated fats as compared to any other nut.
- A. Chestnuts
- B. Pistachios
- C. Pine nuts
- D. Macadamia nuts

_______________ **4.** Red fruits supply ___, which may reduce the risk of some cancers, most notably prostate cancer.
- A. beta-carotene
- B. anthocyanin
- C. lycopene
- D. chlorophyll

_______________ **5.** Some fruits, such as ___, are prone to enzymatic browning.
- A. apples
- B. oranges
- C. blackberries
- D. limes

_______________ **6.** ___ should be ground to facilitate the absorption of their essential fatty acids.
- A. Chia seeds
- B. Flaxseeds
- C. Almonds
- D. Cashews

_______________ **7.** A ___ is a legume found within a thin, netted, tan-colored pod that grows underground.
- A. hazelnut
- B. walnut
- C. pecan
- D. peanut

__________ **8.** A ___ is a fleshy fruit that contains a core of seeds and an edible skin.
A. drupe
B. pome
C. grape
D. berry

__________ **9.** As fruits cook, pectin is released, which naturally ___.
A. thickens the mixture
B. thins the mixture
C. enhances the colors
D. enhances the aromas

__________ **10.** ___ seeds are available in a paste known as tahini.
A. Pumpkin
B. Sesame
C. Poppy
D. Sunflower

__________ **11.** The high fat content in nuts and seeds makes them susceptible to rancidity if not properly stored in a ___ location.
A. cool, dry
B. cool, moist
C. warm, dry
D. warm, moist

__________ **12.** ___ have a higher starch and lower fat content than other nuts and are generally served cooked.
A. Pecans
B. Brazil nuts
C. Chestnuts
D. Macadamia nuts

__________ **13.** ___ are prized for their flavorful zest and juice.
A. Berries
B. Melons
C. Exotic fruits
D. Citrus fruits

__________ **14.** A ___ is a small brown seed from a plant in the mint family.
A. chia seed
B. flaxseed
C. sunflower seed
D. sesame seed

_______________ **15.** A hazelnut, also known as a ___, is a grape-sized nut found in the fuzzy outer husks of the hazel tree.

A. pepita
B. groundnut
C. pignoli
D. filbert

_______________ **16.** ___ walnuts have a more intense flavor than English walnuts.

A. Exotic
B. Tropical
C. Black
D. Brown

_______________ **17.** A tropical fruit is a fruit grown in the humid tropics, where temperatures average ___°F with little variation between winter and summer.

A. 60
B. 80
C. 100
D. 120

_______________ **18.** Nutritionally, berries are a key source of antioxidants as well as ___ and potassium.

A. calcium
B. sodium
C. vitamin A
D. vitamin C

_______________ **19.** Toasting and roasting causes the ___ in nuts and seeds to form a brown exterior surface.

A. proteins and sugars
B. proteins and fats
C. fiber and sugars
D. fiber and fats

_______________ **20.** ___ include fruits such as peaches, nectarines, apricots, avocados, dates, plums, cherries, and olives.

A. Tropical fruits
B. Exotic fruits
C. Drupes
D. Pomes

Completion

_______________ **1.** A(n) ___ is a dry, hard-shelled fruit or seed that contains an inner kernel.

_______________ **2.** A(n) ___ fruit is a fruit that is the result of breeding two or more fruits of the same species that have different characteristics.

______________ **3.** A(n) ___ is a green, bean-shaped nut from a tree.

______________ **4.** A drupe is also known as a(n) ___.

______________ **5.** A(n) ___ fruit is any fruit grown in the moderately humid subtropics where the coldest months average 50°F and summers are hotter than winters.

______________ **6.** ___ is the oxidation of fruits and vegetables, resulting in a brown surface color.

______________ **7.** A(n) ___ is a fruit or unripened ovule of a nonwoody plant.

______________ **8.** A(n) ___ is an oval fruit that has a smooth skin and grows on woody vines in large clusters.

______________ **9.** A pine nut is also known as a(n) ___.

______________ **10.** A pumpkin seed, also known as a(n) ___, is a green seed that comes from a pumpkin plant.

Critical Thinking

Use complete sentences to respond to the following.

1. Compare the nutritional qualities of the six major classifications of fruits.

2. Describe ways in which fruits are used throughout the menu mix.

3. Choose six nuts and/or seeds and describe effective preparation techniques and/or pairings.

4. Describe common preparation techniques used to develop the flavor of nuts and seeds.

5. Summarize the ways in which nuts and seeds can be incorporated into the menu at beverage houses and bar venues and at fine dining and special event venues.

Fruits, Nuts & Seeds on the Menu

Chapter 13

ACTIVITIES

Name ______________________________ **Date** ____________________

Activity 13-1: Sugar Content in Fruits

Fruits are available fresh, frozen, canned, or dried. Fruits that have been processed may contain added ingredients such as sugar.

Access the USDA National Nutrient Database for Standard Reference website at http://ndb.nal.usda.gov/ to determine the amount of sugar, in grams, in each form of the type of fruit listed.

Type	Fresh, Raw (100 g)	Frozen (100 g)	Canned (100 g)	Dried (¼ cup)
Blueberries		*(unsweetened)*	*(heavy syrup)*	*(sweetened)*
Cherries, sour, red		*(unsweetened)*	*(water pack)*	N/A
Cranberries		N/A	*(sauce, sweetened)*	*(sweetened)*
Figs		N/A	*(water pack)*	*(uncooked)*
Grapes, red or green		N/A	*(water pack)*	*(raisins, packed)*
Peaches		*(sweetened)*	*(water pack)*	*(uncooked)*
Pears		N/A	*(water pack)*	*(uncooked)*
Pineapple		*(sweetened)*	*(water pack)*	N/A
Strawberries		*(unsweetened)*	*(heavy syrup)*	N/A
Tangerines		N/A	*(juice pack)*	N/A

Activity 13-2: Fruit, Nut, and Seed Menu Descriptions

Menu descriptions should be clear and appealing while communicating the main ingredients of a dish and the method of preparation. Effective menu descriptions promote clarity, add appeal, and increase the likelihood that the item will be ordered.

Rewrite the following ineffective menu descriptions to make them more appealing and complete.

1. Spiced Nuts and Seeds

2. Chilled Melon-Berry Soup

3. Fruit Salad

4. Granola with Dried Fruits, Nuts, and Seeds

5. Quick Breads with Fruit and Nut Spreads

Menu Descriptions Rubric					
Description Criteria	**Spiced Nuts and Seeds**	**Chilled Melon-Berry Soup**	**Fruit Salad**	**Granola with Dried Fruits, Nuts, and Seeds**	**Quick Breads with Fruit and Nut Spreads**
Accurate representation	1	1	1	1	1
Promotes clarity	1	1	1	1	1
Appealing portrayal	1	1	1	1	1
States unique characteristics	1	1	1	1	1
Conveys preparation techniques	1	1	1	1	1

Scoring: A point is earned for each menu item description that meets the criteria indicated in the rubric for a possible total score of 25 points.

Activity 13-3: Apple Cake Recipe Modification

The purpose of the culinary nutrition recipe-modification process is to increase the nutrient density of a menu item as well as to accommodate guests with dietary considerations such as food allergies. The appearance, texture, and flavor of menu items should also be considered when modifying recipes.

Modify the following apple cake recipe to reduce its calories, fat, and sugar and increase its dietary fiber.

Yield: 12 Slices

Ingredients

2 ea	whole eggs
1 c	vegetable oil
2 c	granulated sugar
1 tsp	vanilla extract
2 c	all-purpose flour
1 tsp	baking soda
2 tsp	ground cinnamon
½ tsp	salt
4 c	Granny Smith apples, peeled and chopped
⅓ c	confectioners' sugar

Preparation

1. Beat the eggs and oil on medium-high speed until creamy.
2. Add the sugar and vanilla and beat until well incorporated. Set aside.
3. Combine the flour, baking soda, cinnamon, and salt in a separate bowl.
4. Slowly add the flour mixture to the egg mixture until well combined.
5. Fold the apples into the prepared batter.
6. Spread the batter into a prepared 9″ × 13″ baking dish.
7. Bake at 350°F for 45 minutes or until a toothpick inserted in the center comes out clean.
8. Allow the cake to cool on a wire rack.
9. Dust with confectioners' sugar.

Culinary Nutrition Recipe-Modification Process

Step 1: Evaluate original recipe for sensory and nutritional qualities.

Step 2: Establish goals for recipe modifications.

Step 3: Identify modifications or substitutions.

Step 4: Determine functions of identified modifications or substitutions.

Step 5: Select appropriate modifications or substitutions.

Step 6: Test modified recipe to evaluate sensory and nutritional qualities.

Activity 13-4: Nut and Seed Butters Lab

Nut and seed butters have unique textures and rich flavors that can be used to develop flavor in both savory and sweet recipes. Although nut and seed butters are high in calories, they contain protein and heart-healthy unsaturated fats.

Objective: To identify nut and seed butters based on appearance, aroma, and flavor.

Materials and Equipment

- Marker
- 5 tasting cups
- Tablespoon
- Nut and seed butters
    - almond butter
    - cashew butter
    - soy nut butter
    - sunflower seed butter
    - tahini (sesame seed butter)
- Blindfold
- 10 tasting spoons

Allergen Alert

The materials used in this activity may contain one or more food allergens.

Procedure

1. Label the bottoms of five of the tasting cups, one type of butter per label.
2. Place 2 tbsp of each butter into the correctly labeled cup.
3. Working in pairs, take turns placing the cups in random order and identifying each butter based on its appearance.
4. Place a blindfold on one partner and have that partner smell and identify each butter based on its aroma.
5. With the blindfold still in place, have the partner taste and identify each butter based on its flavor.
6. Repeat steps 4–5 with the other lab partner wearing the blindfold.

Scoring: A point is earned for each butter correctly identified based on its appearance, aroma, and flavor for a possible total score of 15 points.

Activity 13-5: Nutrient Density of Fruits, Nuts, and Seeds Research

Fruits, nuts, and seeds are nutrient-dense foods rich in antioxidants and phytonutrients, which are also known as phytochemicals. The colors, textures, flavors, and shapes of fruits, nuts, and seeds enliven and add nutrient density to dishes.

Access medical- and nutrition-based websites to complete this activity. Prepare a three- to four-page report and an oral presentation that address the following criteria.

1. Describe three health benefits associated with eating fruits.
2. Explain the benefits of consuming whole fruits over fruit juices.
3. Identify three specific fruits high in antioxidants and phytonutrients.
4. Summarize the health benefits associated with these fruits.
5. Describe three ways to incorporate fruits into the menu.
6. Describe three health benefits associated with eating nuts and/or seeds.
7. Identify three specific nuts and/or seeds high in antioxidants and phytonutrients.
8. Summarize the health benefits associated with these nuts and/or seeds.
9. Describe three ways to incorporate nuts and seeds into the menu.

Nutrition Project Rubric

Content	
Written report	5
Cover page	5
Introduction	5
Quality content	20
Summary	5
Format	
Typed, double spaced	5
Correct spelling	5
Correct grammar	5
Correct punctuation	5
Bibliography	5
Oral Presentation	
Introduction	5
Quality content	10
Summary	5
Delivery	5
Visual aid(s)	10
Total	100

Pastas, Grains & Breads on the Menu

Chapter 14

REVIEW

Name ______________________________ **Date** ____________________

True-False

T F **1.** Rice, wild rice, corn, millet, teff, quinoa, amaranth, and buckwheat all contain gluten.

T F **2.** Most grains must be paired with legumes, nuts, or seeds to make a complete-protein menu item.

T F **3.** Smaller pasta varieties work well in soups because they easily fit on a spoon.

T F **4.** Vegetable-based Asian noodles include lo mein and udon noodles.

T F **5.** The fermentation process is an important step in the development of flavor in yeast breads.

T F **6.** An enriched dough is a yeast dough that is high in sugar and fat.

T F **7.** Most pasta doubles in volume when cooked.

T F **8.** Gums are often added to gluten-free breads to provide the structure that comes from gluten.

T F **9.** The bran and germ of grains can be added to beverages such as smoothies to increase dietary fiber, vitamins, and minerals.

T F **10.** Most grains require processing to make it easier for the body to digest and absorb their nutrients.

T F **11.** Semolina comes from durum wheat and is low in gluten.

T F **12.** When a grain kernel is stripped of both the bran and germ, all the starch is lost.

T F **13.** Baking soda is an acidic chemical leavening agent that reacts to an alkaline dough or batter.

T F **14.** Rice is classified by the size of the grain.

T F **15.** When made with whole grains, breads can be an excellent source of complex carbohydrates and dietary fiber.

Multiple Choice

_______________ **1.** A whole grain is a grain that has only had the ___ removed.
A. husk
B. bran
C. germ
D. endosperm

_______________ **2.** ___ noodles are long, slender noodles made from Japanese yams and do not contain gluten, fat, or calories.
A. La mian
B. Soba
C. Hokkien
D. Shirataki

_______________ **3.** To meet the growing demand for gluten-free menu items, gluten-free pastas made from rice, wild rice, corn, quinoa, ___, or potato flours can be used.
A. wheat
B. rye
C. buckwheat
D. barley

_______________ **4.** In bread dough, ___ acts as a stimulant to the yeast so that fermentation can occur.
A. sugar
B. sodium
C. fat
D. acid

_______________ **5.** Millet, teff, faro, and ___ are considered ancient grains.
A. barley
B. spelt
C. rice
D. corn

_______________ **6.** Common types of ribbon pastas include ___, egg noodles, fettuccine, lasagna, linguine, and spaghetti.
A. campanelle
B. cellantani
C. couscous
D. capellini

_______________ **7.** Common varieties of tube pastas include ___, ditalini, macaroni, manicotti, penne, pipettes, rigatoni, and ziti.
A. rotini
B. capellini
C. cellentani
D. radiatori

____________________ **8.** A ___ grain is a refined grain that has been ground into a fine meal or powder.
A. pearled
B. milled
C. flaked
D. rolled

____________________ **9.** Most grains are cooked by ___.
A. steaming
B. sautéing
C. simmering
D. boiling

____________________ **10.** ___ is the hardest type of wheat.
A. Refined wheat
B. Wheat berry
C. Bulgur wheat
D. Durum wheat

____________________ **11.** Examples of stuffed pastas include ___.
A. ravioli, tortellini, and tortelloni
B. capellini, fettuccine, and linguine
C. farfalle, gemelli, and orecchiette
D. penne, ditalini, and ziti

____________________ **12.** Pasta should be cooked in ___.
A. a steamer
B. a fryer
C. simmering water
D. boiling water

____________________ **13.** ___ noodles are made from buckwheat flour and have a brown-gray color.
A. Cellophane
B. Soba
C. Lo mein
D. Hokkein

____________________ **14.** A yeast starter is a mixture of flour, yeast, ___, and water.
A. sugar
B. salt
C. eggs
D. milk

____________________ **15.** Examples of cereal grains include ___, wheat, oats, barley, corn, rye, triticale, millet, teff, farro, and spelt.
A. buckwheat
B. rice
C. quinoa
D. amaranth

_______________ **16.** ___ is typically made using flour, water, potatoes, and sometimes cheese.
A. A pot sticker
B. Ravioli
C. Tortellini
D. Gnocchi

_______________ **17.** Fresh pasta generally cooks within ___ minutes once it is added to boiling water.
A. 1–2
B. 3–5
C. 6–8
D. 8–11

_______________ **18.** Quick breads are ___ for a short amount of time.
A. mixed
B. proofed
C. fermented
D. kneaded

_______________ **19.** ___ rice is not a true rice, but comes from an aquatic grass and has a rich, nutty flavor.
A. Arborio
B. Basmati
C. Wild
D. Red

_______________ **20.** ___ provides leavening in quick breads twice during the baking process.
A. Yeast
B. Baking soda
C. Baking powder
D. Double-acting baking powder

_______________ **21.** Hard rolls, baguettes, and rye breads are made from ___ doughs.
A. lean
B. rich
C. enriched
D. rolled-in

_______________ **22.** ___ is the process of pushing and folding dough until it is smooth and elastic.
A. Docking
B. Proofing
C. Kneading
D. Punching

_______________ **23.** ___ oats are oat groats that have been toasted and cut into small pieces.
A. Steel-cut
B. Old-fashioned
C. Rolled
D. Milled

________________ **24.** ___ is the process of letting yeast dough rise in a warm (85°F) and moist (80% humidity) environment until the dough doubles in size.
A. Scoring
B. Proofing
C. Docking
D. Punching

________________ **25.** Quick breads include items such as pancakes, waffles, biscuits, scones, muffins, coffee cakes, and ___.
A. focaccias
B. baguettes
C. cornbreads
D. croissants

________________ **26.** When the temperature of yeast dough reaches ___°F, the yeast dies and there is no further expansion.
A. 80
B. 100
C. 120
D. 140

________________ **27.** ___ is a gluten-free pseudocereal with a small, yellow, black-flecked seed and is high in protein.
A. Amaranth
B. Buckwheat
C. Millet
D. Farro

________________ **28.** The flour used to make pasta can come from any grain, but ___ flour is the most common.
A. semolina
B. buckwheat
C. rice
D. oat

________________ **29.** ___ is classified as a complete protein.
A. Amaranth
B. Teff
C. Spelt
D. Quinoa

________________ **30.** When wheat flour is mixed with a liquid, usually water or milk, glutenin and gliadin join together to form ___.
A. gluten
B. glucose
C. lactose
D. glycogen

Completion

______________ **1.** A(n) ___ is a seed that is classified as a grain but is derived from broadleaf plants instead of grasses.

______________ **2.** A(n) ___ is any ingredient that causes a baked product to rise by the action of air, steam, chemicals, or yeast.

______________ **3.** ___ is the process of making small holes in dough before it is baked to allow steam to escape and to promote even baking.

______________ **4.** ___ is a hybrid cereal grain made by crossing rye and wheat.

______________ **5.** Pasta is cooked until it becomes ___, meaning "to the tooth."

______________ **6.** ___ is a microscopic, single-celled fungus that releases carbon dioxide and alcohol through a process called fermentation when provided with food (sugar) in a warm, moist environment.

______________ **7.** ___ pasta is pasta that has been extruded into complex shapes such as corkscrews, bowties, shells, flowers, or stars.

______________ **8.** A(n) ___ grain is a grain that has been processed to remove the bran, germ, or both.

______________ **9.** ___ is the breakdown of carbohydrates due to the actions of bacteria, microorganisms, or yeast.

______________ **10.** A(n) ___ is the edible fruit, in the form of a kernel or seed, of a grass.

______________ **11.** ___ pasta is pasta that has been formed by hand or machine to hold a filling.

______________ **12.** ___ is a gluten-free pseudocereal with a dark, three-cornered seed.

______________ **13.** ___ yeast is yeast that has approximately 70% moisture and is available in 1 lb cakes or blocks.

______________ **14.** ___ noodles, also known as glass noodles, are made from mung bean starch and are fairly transparent.

______________ **15.** ___ are oats that have been steamed and flattened into small flakes.

______________ **16.** ___ is a gluten-free cereal grain cultivated from an annual grass that bears kernels on large woody cobs called ears.

______________ **17.** ___ is a type of protein found in grains such as wheat, rye, and barley.

______________ **18.** Baking soda produces ___ when it comes in contact with an acidic ingredient.

______________ **19.** ___ is a traditional Italian dish typically made by sautéing and then simmering Arborio rice.

______________ **20.** ___ is a gluten-free ancient cereal grain and is the smallest grain in the world.

Critical Thinking

Use complete sentences to respond to the following.

1. Describe how to develop flavor in pasta dishes using healthy preparation techniques.

2. Describe healthy ways in which pasta can be used throughout the menu mix.

3. Explain how processing affects the nutritional qualities of grains.

4. Identify benefits of using grains in beverages, appetizers, soups, salads, entrées, sides, and desserts.

5. Compare yeast breads and quick breads in terms of preparation.

6. Describe healthy ways in which breads can be used throughout the menu mix.

Pastas, Grains & Breads on the Menu

Chapter 14

ACTIVITIES

Name ______________________________ **Date** ______________

Activity 14-1: Pasta Shapes and Sizes

Pastas come in many shapes and sizes and thus can enhance plated presentations. The shape and size of the pasta chosen for a dish is often determined by how the sauce will adhere to the pasta and how the pasta will complement the dish.

Match each type of pasta to its image.

______________ **1.** Fettuccine

______________ **2.** Linguine

______________ **3.** Lasagna

______________ **4.** Campanelle

______________ **5.** Farfalle

______________ **6.** Fiori

______________ **7.** Orecchiette

______________ **8.** Orzo

______________ **9.** Rotini

______________ **10.** Cellentani

______________ **11.** Ditalini

______________ **12.** Manicotti

______________ **13.** Penne

______________ **14.** Tortellini

______________ **15.** Tortelloni

Ⓐ

Ⓑ

Ⓒ

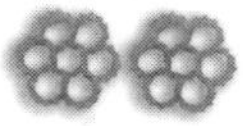
Ⓓ

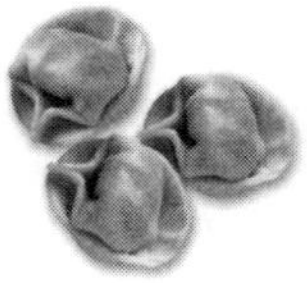
Ⓔ

Ⓕ

Ⓖ

Ⓗ

Ⓘ

Ⓙ

Ⓚ

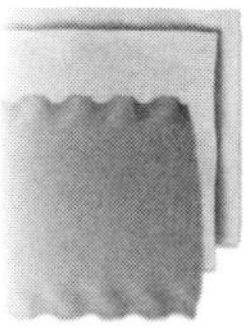
Ⓛ

Ⓜ

Ⓝ

Ⓞ

Barilla America, Inc.

Activity 14-2: Gluten-Free and Ancient Grains

Gluten-free and ancient grains are often perceived as healthier menu options than refined grains. These grains add value to a menu because of their nutritional benefits, which include protein, omega-3 fatty acids, and antioxidants.

Identify the item from each of the following pairs that is gluten-free.

____________________ **1.** amaranth / rye

____________________ **2.** barley / millet

____________________ **3.** wheat berries / corn

____________________ **4.** wild rice / semolina

____________________ **5.** teff / bulgur

____________________ **6.** buckwheat / spelt

____________________ **7.** triticale / rice

____________________ **8.** quinoa / farro

Identify the item from each of the following pairs that is an ancient grain.

____________________ **9.** teff / corn

____________________ **10.** rice / amaranth

____________________ **11.** bulgur / buckwheat

____________________ **12.** quinoa / rye

____________________ **13.** farro / barley

____________________ **14.** triticale / spelt

Activity 14-3: Rotini with Creamy Seafood Sauce Recipe Modification

The purpose of the culinary nutrition recipe-modification process is to increase the nutrient density of a menu item as well as to accommodate guests with dietary considerations such as food allergies. The appearance, texture, and flavor of menu items should also be considered when modifying recipes.

Modify the following rotini recipe to reduce its calories, fat, cholesterol, and sodium; increase its dietary fiber; and accommodate guests with seafood allergies.

Yield: 8 servings, ⅔ cup each

Ingredients

1 lb	rotini, dried
2 c	heavy cream
1 tbsp	basil, chopped
1 tbsp	thyme, chopped
2 tsp	kosher salt
2 tsp	red pepper flakes
1 c	green onions, chopped
1 c	parsley, chopped
⅔ lb	shrimp, peeled and deveined
⅔ lb	scallops
½ c	Monterey Jack cheese, shredded
½ c	Parmesan cheese, shredded

Preparation

1. Bring 10 cups of water to a boil.
2. Add the rotini and cook until al dente. Drain and set aside.
3. Pour cream into a skillet and bring to a simmer over medium heat while stirring constantly.
4. Add basil, thyme, salt, pepper flakes, onions, and parsley. Stir to combine.
5. Simmer the cream mixture until thickened (approximately 7–8 minutes).
6. Stir in the shrimp and scallops and cook until the seafood is firm and opaque.
7. Add the cheeses and stir to combine.
8. Add the pasta to the sauce, stir to combine, and serve.

Culinary Nutrition Recipe-Modification Process

Step 1: Evaluate original recipe for sensory and nutritional qualities.

Step 2: Establish goals for recipe modifications.

Step 3: Identify modifications or substitutions.

Step 4: Determine functions of identified modifications or substitutions.

Step 5: Select appropriate modifications or substitutions.

Step 6: Test modified recipe to evaluate sensory and nutritional qualities.

Activity 14-4: Pasta, Grain, and Bread Menu Descriptions

Menu descriptions should be clear and appealing while communicating the main ingredients of a dish and the method of preparation. Effective menu descriptions promote clarity, add appeal, and increase the likelihood that the item will be ordered.

Rewrite the following ineffective menu descriptions to make them more appealing and complete.

1. Bruschetta

2. Tabbouleh Salad

3. Pasta Primavera

4. Ancient-Grain Risotto

5. Mini-Muffin Assortment

Menu Descriptions Rubric					
Description Criteria	**Bruschetta**	**Tabbouleh Salad**	**Pasta Primavera**	**Ancient-Grain Risotto**	**Mini-Muffin Assortment**
Accurate representation	1	1	1	1	1
Promotes clarity	1	1	1	1	1
Appealing portrayal	1	1	1	1	1
States unique characteristics	1	1	1	1	1
Conveys preparation techniques	1	1	1	1	1

Scoring: A point is earned for each menu item description that meets the criteria indicated in the rubric for a possible total score of 25 points.

Activity 14-5: Pastas Made with Various Flours Lab

Objective: To evaluate pastas made with various flours in terms of color, texture, and flavor.

Materials and Equipment

- Labels
- Marker
- 4 bowls
- 4 stockpots
- Water
- 8 oz of each pasta
  - refined semolina flour pasta
  - whole wheat flour pasta
  - rice flour pasta
  - quinoa flour pasta
- Spoon
- Strainer
- Tasting forks

Allergen Alert

The materials used in this activity may contain one or more food allergens.

Procedure

1. Label the first bowl "refined pasta," the second bowl "whole wheat pasta," the third bowl "rice pasta," and the fourth bowl "quinoa pasta."
2. Place four labels that are identical to those in step 1 on each of the four stockpots.
3. Fill each stockpot with approximately 2–3 quarts of water and bring to a boil.
4. Add each type of pasta to its corresponding pot and cook until al dente while stirring occasionally.
5. Drain each pot of pasta, placing the pasta in its corresponding bowl after it is drained.
6. Evaluate each type of pasta in terms of color, texture, and flavor.

Respond to the following.

1. Compare the color of each type of pasta.

2. Compare the texture of each type of pasta.

3. Compare the flavor of each type of pasta.

Activity 14-6: Pasta, Grain, and Bread Research

Pastas, grains, and breads are the primary sources of carbohydrates in a balanced diet. They are popular around the globe and come in a variety of forms.

Access an article from a medical-based website, a nutrition journal, and a company that sells whole grains. Prepare a three- to four-page report and an oral presentation that address the following criteria.

1. Explain why carbohydrates are necessary for health.
2. Differentiate between simple and complex carbohydrates.
3. Identify five foods that are simple carbohydrates and five foods that are complex carbohydrates.
4. Describe the health benefits of consuming whole grains.
5. Identify three types of whole grains, including one ancient grain and one gluten-free grain.
6. Describe how these grains have been used historically.
7. Compare these grains in terms of fat, protein, dietary fiber, and key vitamins and minerals.
8. Describe how these grains can be incorporated into the menu mix.

Nutrition Project Rubric

Content	
Written report	5
Cover page	5
Introduction	5
Quality content	20
Summary	5
Format	
Typed, double spaced	5
Correct spelling	5
Correct grammar	5
Correct punctuation	5
Bibliography	5
Oral Presentation	
Introduction	5
Quality content	10
Summary	5
Delivery	5
Visual aid(s)	10
Total	100

Desserts on the Menu — Chapter 15 REVIEW

Name ______________________________ Date ____________________

True-False

T F **1.** Low-fat milk can be used in place of cream in ice cream, but it produces a grainier texture.

T F **2.** Most pastry doughs are high in saturated fat.

T F **3.** Butter should be substituted for oil to reduce saturated fats.

T F **4.** Many confections are high in empty calories.

T F **5.** Preparing desserts in advance allows the ingredients to disperse and the flavors to intensify.

T F **6.** Cookie dough typically contains more fat and less moisture than cake batter.

T F **7.** Dessert cheeses are served chilled for optimum flavor.

T F **8.** Simple syrup is made by heating a mixture of cream and sugar until the sugar dissolves.

T F **9.** Tarts can have a double crust.

T F **10.** Because aromatic ingredients amplify flavor, the amount of sugar added to a dessert can often be reduced.

T F **11.** Dessert menus should emphasize seasonal ingredients that are at their peak of flavor.

T F **12.** Dry-heat cooking techniques naturally concentrate the sugars in fruits to create depth of flavor.

T F **13.** Sherbets are higher in fat and calories than ice cream.

T F **14.** When preparing frozen yogurt, the freezing process destroys most of the beneficial live bacteria in the yogurt.

T F **15.** When serving reduced portion sizes, the size of the dessert plate should also be reduced.

Multiple Choice

_______________ **1.** A ___ is a light and foamy custard prepared by whisking egg yolks, sugar, and white wine over a double boiler into a silky foam.
A. pavlova
B. meringue
C. mousse
D. sabayon

_______________ **2.** ___ are generally topped with a thick biscuit crust and sprinkled with sugar.
A. Crisps
B. Cobblers
C. Tarts
D. Pies

_______________ **3.** ___ is a light and refreshing fat-free and dairy-free dessert that is relatively low in calories but high in sugar.
A. Ice cream
B. Frozen yogurt
C. Sorbet
D. Sherbet

_______________ **4.** ___ are made from a batter of eggs, milk, and flour and are often filled with fruits or pastry cream.
A. Tarts
B. Confections
C. Tortes
D. Crêpes

_______________ **5.** Custards do not contain ___.
A. cream
B. flour
C. sugar
D. eggs

_______________ **6.** The main difference between pies and tarts is their ___.
A. depth
B. crust ingredients
C. fillings
D. flavor profiles

_______________ **7.** A crisp cookie contains a high percentage of ___.
A. sugar
B. flour
C. butter
D. egg

________________ **8.** A meringue is a mixture of ___ and sugar.
- A. egg whites
- B. egg yolks
- C. cream
- D. butter

________________ **9.** Cobblers and crisps do not have ___.
- A. a topping
- B. a bottom crust
- C. fillings
- D. fruit

________________ **10.** ___ is a frozen dessert made from cream, butterfat, sugar, and sometimes eggs.
- A. Sorbet
- B. Sherbet
- C. Ice cream
- D. Frozen yogurt

Completion

________________ **1.** ___ is a composed frozen dessert made with a sponge cake soaked in a liqueur-flavored syrup that is topped with ice cream and covered with Italian meringue.

________________ **2.** A(n) ___ is a sugar-based coating often spread on the outside of and between the layers of a baked product.

________________ **3.** ___ is a frozen dessert made from yogurt, sugar, and sometimes gelatin.

________________ **4.** A(n) ___ is a small sugar-based food item such as a candy or a chocolate.

________________ **5.** A(n) ___ is a rich cake that is usually layered with creams, jams, or fruits and is iced with a sweet icing or glazed and garnished.

________________ **6.** ___ is a crisp meringue that is filled with whipped cream and fruit.

________________ **7.** A(n) ___ is a rich and airy foam that has been lightened with whipped cream, whipped pasteurized egg whites, or both.

________________ **8.** A(n) ___ is a baked custard that typically has a graham cracker or cookie crust.

________________ **9.** A(n) ___ is a frozen dessert made by combining puréed fresh fruit, simple syrup, and milk.

________________ **10.** A(n) ___ is a French pancake that is light and very thin.

Critical Thinking

Use complete sentences to respond to the following.

1. Describe how to effectively plate desserts.

2. Summarize the ways that beverages, cheeses, fruits, nuts and seeds, and breads can be served as desserts and include nutritional concerns that may accompany these items.

3. Explain how signature desserts, tasting portions, and dessert trays can be used to promote desserts.

Desserts on the Menu Chapter 15 ACTIVITIES

Name ______________________________ Date ____________________

Activity 15-1: Dessert Components

Plated desserts often contain a variety of components that can make them high in calories and fats. To make more nutrient-dense desserts, dessert components should be limited in saturated fats.

Circle the two components in each dessert that contain the lowest amount of saturated fat.

1. chocolate-banana cake / butterscotch sauce / fresh banana-berry salsa / toasted slivered almonds

2. baked vanilla meringue / whipped cream / strawberry-orange basil sorbet / shaved dark chocolate

3. lemon caramel pound cake / lemon-almond ice milk / citrus coulis / white chocolate almond cookies

4. baked figs / maple-cinnamon Neufchâtel / pecan shortbread / buttered rum sauce

5. honey-Greek yogurt / strawberry-mint relish / ladyfingers / chocolate sauce

6. spice cake / mango-pineapple salad / toffee sauce / roasted macadamia nuts

7. chocolate torte / vanilla bean frozen yogurt / grilled pears / amaretto crème anglaise

8. pumpkin tofu purée / pumpkin seed brittle / brandy-infused golden raisins / ganache drizzle

9. angel food cake / lemon custard / berry medley / toasted coconut

10. gingered fruit salad / crème fraîche / agave nectar / gingersnap cookies

Activity 15-2: Dessert Modification Strategies

Effective dessert modifications should reduce calories, fats, and/or sugar while still producing a flavorful dessert.

Match the letter that represents the missing word in each of the following statements.

______________ **1.** Reduce sugars by ___ to two-thirds.

______________ **2.** Replace ___ sugars with natural sugars such as honey, maple syrup, and agave nectar.

______________ **3.** Replace refined ___ with whole grain varieties.

______________ **4.** ___ low-fat dairy products for full-fat varieties.

______________ **5.** Replace butter and shortening with ___ oils.

______________ **6.** Replace ___ with applesauce, banana purée, or prune purée.

______________ **7.** Use egg substitutes or egg ___ instead of whole eggs.

______________ **8.** Incorporate fresh, seasonal fruits; avoid fruits ___ in syrup.

______________ **9.** Use flavor ___ such as citrus zest, vanilla beans, and spices.

______________ **10.** Add liqueurs and wines to ___ liquids and sauces.

______________ **11.** Roast nuts and use them ___.

______________ **12.** Replace ___ with healthier cooking techniques.

A. Whites
B. Sparingly
C. Half
D. Canned
E. Substitute
F. Refined
G. Oil
H. Frying
I. Flours
J. Poaching
K. Unsaturated
L. Enhancers

Activity 15-3: Chocolate Panna Cotta Recipe Modification

The purpose of the culinary nutrition recipe-modification process is to increase the nutrient density of a menu item as well as to accommodate guests with dietary considerations such as food allergies. The appearance, texture, and flavor of menu items should also be considered when modifying recipes.

Modify the following chocolate panna cotta recipe to reduce its calories, fat, and sugar; increase its vitamins, minerals, and antioxidants; and accommodate lactose-intolerant guests.

Yield: 4 servings

Ingredients

2 tsp	powdered gelatin
2 tbsp	water
9 fl oz	heavy cream
3 fl oz	whole milk
2.5 oz	granulated sugar
3 oz	milk chocolate, chopped
¼ tsp	vanilla extract
1 c	cherries, canned in syrup, drained

Preparation

1. Place the gelatin in a small bowl and mix in the water. Allow to rest 10 minutes.
2. Add the cream, milk, and sugar to a saucepan and cook over medium heat until the mixture begins to boil slightly. Remove from heat.
3. Add the chocolate and vanilla and stir until the chocolate melts.
4. Add the gelatin and stir until well combined.
5. Divide the mixture among four ramekins and refrigerate for 4 hours or overnight.
6. Unmold the panna cottas onto four plates.
7. Top each panna cotta with ¼ cup sweetened cherries.

Culinary Nutrition Recipe-Modification Process

Step 1: Evaluate original recipe for sensory and nutritional qualities.

Step 2: Establish goals for recipe modifications.

Step 3: Identify modifications or substitutions.

Step 4: Determine functions of identified modifications or substitutions.

Step 5: Select appropriate modifications or substitutions.

Step 6: Test modified recipe to evaluate sensory and nutritional qualities.

Activity 15-4: Dessert Menu Descriptions

Menu descriptions should be clear and appealing while communicating the main ingredients of a dish and the method of preparation. Effective menu descriptions promote clarity, add appeal, and increase the likelihood that the item will be ordered.

Rewrite the following ineffective menu descriptions to make them more appealing and complete.

1. Tropical Fruit Trifle

2. Chocolate Rice Pudding

3. Banana Nut-Butter Crêpes

4. Frozen Yogurt Parfait

5. Glazed Grilled Fruits

Menu Descriptions Rubric					
Description Criteria	**Tropical Fruit Trifle**	**Chocolate Rice Pudding**	**Banana Nut-Butter Crêpes**	**Frozen Yogurt Parfait**	**Glazed Grilled Fruits**
Accurate representation	1	1	1	1	1
Promotes clarity	1	1	1	1	1
Appealing portrayal	1	1	1	1	1
States unique characteristics	1	1	1	1	1
Conveys preparation techniques	1	1	1	1	1

Scoring: A point is earned for each menu item description that meets the criteria indicated in the rubric for a possible total score of 25 points.

Activity 15-5: Fat Replacers in Cakes Lab

Fat replacers used in baked goods include puréed pumpkin, prunes, legumes, silken tofu, mashed bananas, applesauce, and yogurt. Fat replacers typically reduce fats and increase nutrient density.

Objective: To evaluate cakes made with butter, yogurt, and applesauce in terms of texture and flavor.

Materials and Equipment

- 6 cake pans (9-inch)
- Nonstick cooking spray
- Measuring cup
- Measuring spoons
- 6 mixing bowls
- Mixer
- Sifter
- Rubber spatula
- 3 plates
- Tasting forks
- Ingredients for butter cake, yogurt cake, and applesauce cake: See chart

Allergen Alert

The materials used in this activity may contain one or more food allergens.

Butter Cake Ingredients	Yogurt Cake Ingredients	Applesauce Cake Ingredients
• 1½ c butter, room temperature • 2½ c granulated sugar • 5 whole eggs • 1 tsp pure vanilla extract • 3 c all-purpose flour • ¾ tsp baking powder • ¼ tsp salt • 1 c whole milk	• 1½ c plain, low-fat yogurt • 2½ c granulated sugar • 5 whole eggs • 1 tsp pure vanilla extract • 3 c all-purpose flour • ¾ tsp baking powder • ¼ tsp salt • 1 c whole milk	• 1½ c applesauce • 2½ c granulated sugar • 5 whole eggs • 1 tsp pure vanilla extract • 3 c all-purpose flour • ¾ tsp baking powder • ¼ tsp salt • 1 c whole milk

Preparation

1. Coat two 9-inch cake pans with nonstick cooking spray and set aside.
2. Add the butter and sugar in a mixing bowl and beat well.
3. Add eggs one at a time, beating well after each egg is added. Mix in the vanilla.
4. Sift the flour, baking powder, and salt into a separate bowl.
5. Add the flour mixture to the sugar mixture in one-third increments, mixing until combined and scraping down the sides and bottom of the bowl with a spatula as needed.
6. Add the milk and mix until incorporated.
7. Pour the batter into the prepared cake pans and bake at 350°F for 25 minutes or until a toothpick inserted in the center of each cake comes out clean.
8. Allow the cakes to cool in the pans for 10 minutes and then remove them from the pans to cool completely.
9. Repeat steps 1–8, substituting plain, low-fat yogurt for the butter.
10. Repeat steps 1–8, substituting applesauce for the butter.
11. Place a slice of each cake on a plate and evaluate each slice in terms of texture and flavor.

Respond to the following.

1. Compare each slice of cake in terms of texture and flavor.

Activity 15-6: Coconut Oil Research

Preliminary studies reveal that coconut oil may have a variety of health benefits. However, moderate consumption of coconut oil is recommended because it is high in calories.

Access an article from a medical-based website, a nutrition journal, and a company that sells coconut oil. Prepare a three- to four-page report and an oral presentation that address the following criteria.

1. Differentiate between refined and unrefined coconut oil.
2. Describe three potential health benefits associated with the use of coconut oil.
3. Identify any dietary concerns associated with coconut oil.
4. Compare coconut oil and olive oil in terms of calories, fat, and saturated fat.
5. Explain how the saturated fats in coconut oil differ from the saturated fats in animal-based foods.
6. Explain why coconut oil is effective in vegetarian baking.
7. Explain how this research will affect your selection of coconut oil in the future.

Nutrition Project Rubric

Content	
Written report	5
Cover page	5
Introduction	5
Quality content	20
Summary	5
Format	
Typed, double spaced	5
Correct spelling	5
Correct grammar	5
Correct punctuation	5
Bibliography	5
Oral Presentation	
Introduction	5
Quality content	10
Summary	5
Delivery	5
Visual aid(s)	10
Total	100